Principles of Atomic Physics

Principles of Atomic Physics

Titus Mason

Willford Press
www.willfordpress.com

Published by Willford Press,
118-35 Queens Blvd., Suite 400,
Forest Hills, NY 11375, USA

ISBN: 978-1-68285-960-5

Cataloging-in-Publication Data

Principles of atomic physics / Titus Mason.
p. cm.
Includes bibliographical references and index.
ISBN 978-1-68285-960-5
1. Nuclear physics. 2. Atoms. 3. Physics. 4. Nuclear physicists.
I. Mason, Titus.
QC173 .P75 2020
539.7--dc23

For information on all Willford Press publications
visit our website at www.willfordpress.com

Willford Press

Contents

Preface ..... **VII**

Chapter 1 **Atomic Physics: An Introduction** **1**
- i. Atoms 1
- ii. Atomic Physics 23

Chapter 2 **Models of Atomic Structure** **25**
- i. Dalton's Atomic Model 25
- ii. Thomson's Plum Pudding Model 28
- iii. Rutherford Model 29
- iv. Bohr Model of Atom 31
- v. Sommerfeld Model of the Atom 33
- vi. Quantum Mechanical Model of Atom 36

Chapter 3 **The Hydrogen Atom** **48**
- i. Isotopes of Hydrogen 49
- ii. Bohr Model of Hydrogen Atom 61
- iii. Quantum Mechanical Model of Hydrogen Atom 64
- iv. Hydrogen Spectrum 71
- v. Hydrogen-like Atoms 72

Chapter 4 **Multi-electron Atoms** **80**
- i. Helium Atom 94

Chapter 5 **Atomic Spectroscopy** **100**
- i. Atomic Spectra 120
- ii. Alkali Spectra 124
- iii. Fine Structure 138
- iv. Hyperfine Structure 148
- v. Einstein Coefficients 152
- vi. Interaction of an Atom with Radiation 159

Chapter 6 **Spin-orbit Coupling** **163**
- i. LS Coupling 165
- ii. JJ Coupling 171

Chapter 7 **Atoms in External Fields** **174**
i. Zeeman Effect 174
ii. Paschen-Back Effect 179
iii. Stark Effect 181

Permissions

Index

Preface

The field of physics which studies atoms as an atomic nucleus and an isolated system of electrons is known as atomic physics. Its fundamental concern is the arrangement of electrons around the nucleus and the mechanisms through which these arrangements change. Both neutral atoms and ions are studied under this discipline. Atomic physics undertakes atoms to be in isolation, and studies them accordingly. The processes of ionization and excitation by photons or collisions with atomic particles are also dealt within this field. The underlying theory in plasma physics and atmospheric physics has been provided by atomic physics. This book discusses the fundamentals as well as modern approaches of this field. Coherent flow of topics, student-friendly language and extensive use of examples make this book an invaluable source of knowledge. This book is an essential guide for both academicians and those who wish to pursue this discipline further.

A foreword of all chapters of the book is provided below:

Chapter 1 - The smallest constituent unit of matter which makes up a chemical element is called an atom. The domain of physics which focuses on the study of atoms as an isolated system of electrons and an atomic nucleus is known as atomic physics. This is an introductory chapter which will provide a brief introduction to all the significant aspects of atomic physics; **Chapter 2** - A model which represents what the structure of an atom could look like on the basis of the knowledge of the behavior of atoms is known as model of atomic structure. There are a number of models which have been presented by different physicists such as Dalton, Thomson, Rutherford and Bohr. The topics elaborated in this chapter will help in gaining a better perspective about the models put forward by these physicists; **Chapter 3** - The atom of the chemical element hydrogen is called hydrogen atom. In its electrically neutral form, it contains a single proton and electron. Some of its isotopes are Deuterium and Tritium. This chapter has been carefully written to provide an easy understanding of the varied facets of hydrogen atom as well as its different models; **Chapter 4** - The atoms which have more than one electron are called multi-electron atoms. The presence of more than one electron in the atom leads to phenomena like shielding. This chapter closely examines the key concepts of multi electron atoms such as helium to provide an extensive understanding of the subject; **Chapter 5** - The field of study which focuses on the electromagnetic radiation which is absorbed and emitted by atoms is called atomic spectroscopy. Some of the focus areas of this field are Einstein coefficients and interaction of an atom with radiation. The diverse aspects of atomic spectroscopy such as atomic spectra and alkali spectra have been thoroughly discussed in this chapter; **Chapter 6** - The relativistic interaction of a particle's spin with its motion inside a potential is referred to as spin-orbit coupling. Some of the different types of coupling schemes are JJ coupling and RS coupling. The topics elaborated in this chapter will help in gaining a better perspective about these types of spin-orbit coupling; **Chapter 7** - There are a number of phenomena associated with the atoms in external electromagnetic fields such as Zeeman Effect, Paschen-Back Effect and Stark Effect. The topics elaborated in this chapter will help in gaining a better perspective about the behavior and phenomena related to the atoms in external fields.

At the end, I would like to thank all the people associated with this book devoting their precious time and providing their valuable contributions to this book. I would also like to express my gratitude to my fellow colleagues who encouraged me throughout the process.

Titus Mason

1

Atomic Physics: An Introduction

The smallest constituent unit of matter which makes up a chemical element is called an atom. The domain of physics which focuses on the study of atoms as an isolated system of electrons and an atomic nucleus is known as atomic physics. This is an introductory chapter which will provide a brief introduction to all the significant aspects of atomic physics.

Atoms

An atom is a submicroscopic structure found in all ordinary matter. Originally the atom was believed to be the smallest possible indivisible particle of matter. Later, atoms were found to be composed of even smaller subatomic particles. Consisting of a positively charged nucleus surrounded by a cloud of negatively charged electrons, atoms exhibit the duality of positivity and negative that is characteristic of all existing beings. Atoms are the fundamental building blocks of matter. They can be classed into elements, and combine in definite ratios to form compounds through ionic or covalent bonding. In chemical reactions they are neither created nor destroyed, and are said to be conserved.

Most matter consists of an agglomeration of molecules, which can be separated relatively easily. Molecules, in turn, are composed of atoms joined by chemical bonds that are more difficult to break. Each individual atom consists of smaller particles—namely, electrons and nuclei. These particles are electrically charged, and the electric forces on the charge are responsible for holding the atom together. Attempts to separate these smaller constituent particles require ever-increasing amounts of energy and result in the creation of new subatomic particles, many of which are charged.

An atom consists largely of empty space. The nucleus is the positively charged centre of an atom and contains most of its mass. It is composed of protons, which have a positive charge, and neutrons, which have no charge. Protons, neutrons, and the electrons surrounding them are long-lived particles present in all ordinary, naturally occurring atoms. Other subatomic particles may be found in association with these three types of particles. They can be created only with the addition of enormous amounts of energy, however, and are very short-lived.

All atoms are roughly the same size, whether they have 3 or 90 electrons. Approximately 50 million atoms of solid matter lined up in a row would measure 1 cm (0.4 inch). A convenient unit of length for measuring atomic sizes is the angstrom (Å), defined as 10^{-10} metre. The radius of an atom measures 1–2 Å. Compared with the overall size of the atom, the nucleus is even more minutes. It is in the same proportion to the atom as a marble is to a football field. In volume the nucleus takes up only 10^{-14} metres of the space in the atom—i.e., 1 part in 100,000. A convenient unit of length for measuring nuclear sizes is the femtometre (fm), which equals 10^{-15} metre. The diameter of a nucleus depends on the number of particles it contains and ranges from about 4 fm for a light nucleus such as carbon to 15 fm for a heavy nucleus such as lead. In spite of the small size of the nucleus, virtually all the mass of the atom is concentrated there. The protons are massive, positively charged particles, whereas the neutrons have no charge and are slightly more massive than the protons. The fact is that nuclei can have anywhere from 1 to nearly 300 protons and neutrons accounts for their wide variation in mass. The lightest nucleus that of hydrogen, is 1,836 times more massive than an electron, while heavy nuclei are nearly 500,000 times more massive.

Basic Properties

Atomic Number

The single most important characteristic of an atom is its atomic number (usually denoted by the letter *Z*), which is defined as the number of units of positive charge (protons) in the nucleus. For example, if an atom has a *Z* of 6, it is carbon, while a *Z* of 92 corresponds to uranium. A neutral atom has an equal number of protons and electrons so that the positive and negative charges exactly balance. Since it is the electrons that determine how one atom interacts with another, in the end it is the number of protons in the nucleus that determines the chemical properties of an atom.

Atomic Mass and Isotopes

The number of neutrons in a nucleus affects the mass of the atom but not its chemical properties. Thus, a nucleus with six protons and six neutrons will have the same chemical properties as a nucleus with six protons and eight neutrons, although the two masses will be different. Nuclei with the same number of protons but different numbers of neutrons are said to be isotopes of each other. All chemical elements have many isotopes.

Isotopes of hydrogen.

The image shows the three isotopes of the element hydrogen. All three forms have one proton (pink) and one electron (dark green) but differ in the number of neutrons (gray) in the nucleus. Protium, or ordinary hydrogen (top), has no neutrons. Deuterium, or hydrogen-2 (bottom left), has one neutron. Tritium, or hydrogen-3 (bottom right), has two neutrons.

It is usual to characterize different isotopes by giving the sum of the number of protons and neutrons in the nucleus—a quantity called the atomic mass number. In the above example, the first atom would be called carbon-12 or ^{12}C (because it has six protons and six neutrons), while the second would be carbon-14 or ^{14}C.

The mass of atoms is measured in terms of the atomic mass unit, which is defined to be $^{1}/_{12}$ of the mass of an atom of carbon-12, or $1.660538921 \times 10^{-24}$ gram. The mass of an atom consists of the mass of the nucleus plus that of the electrons, so the atomic mass unit is not exactly the same as the mass of the proton or neutron.

Electron

Charge, Mass and Spin

Scientists have known since the late 19th century that the electron has a negative electric charge. The value of this charge was first measured by the American physicist Robert Millikan between 1909 and 1910. In Millikan's oil-drop experiment, he suspended tiny oil drops in a chamber containing an oil mist. By measuring the rate of fall of the oil drops, he was able to determine their weight. Oil drops that had an electric charge (acquired, for example, by friction when moving through the air) could then be slowed down or stopped by applying an electric force. By comparing applied electric force with changes in motion, Millikan was able to determine the electric charge on each drop. After he had measured many drops, he found that the charges on all of them were simple multiples of a single number. This basic unit of charge was the charge on the electron, and the different charges on the oil drops corresponded to those having 2, 3, 4,... extra electrons on them. The charge on the electron is now accepted to be $1.602176565 \times 10^{-19}$ coulomb. For this work Millikan was awarded the Nobel Prize for Physics in 1923.

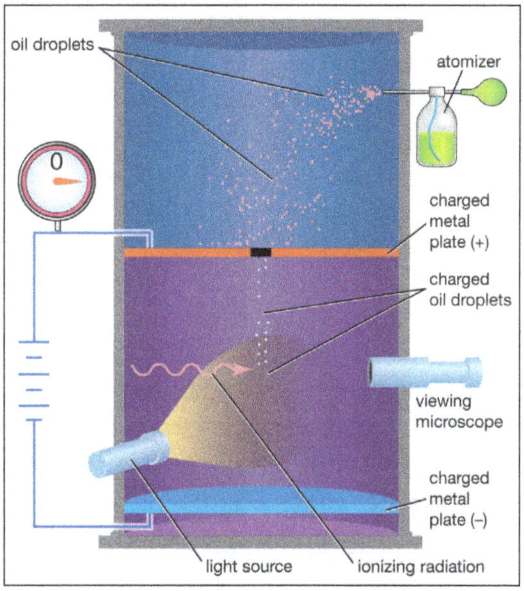

Millikan oil-drop experiment.

Between 1909 and 1910, the American physicist Robert Millikan conducted a series of oil-drop experiments. By comparing applied electric force with changes in the motion of the oil drops, he was able to determine the electric charge on each drop. He found that all of the drops had charges that were simple multiples of a single number, the fundamental charge of the electron.

The charge on the proton is equal in magnitude to that on the electron but opposite in sign—that is, the proton has a positive charge. Because opposite electric charges attract each other, there is an attractive force between electrons and protons. This force is what keeps electrons in orbit around the nucleus, something like the way that gravity keeps Earth in orbit around the Sun.

The electron has a mass of about $9.109382911 \times 10^{-28}$ gram. The mass of a proton or neutron is about 1,836 times larger. This explains why the mass of an atom is primarily determined by the mass of the protons and neutrons in the nucleus.

The electron has other intrinsic properties. One of these is called spin. The electron can be pictured as being something like Earth, spinning around an axis of rotation. In fact, most elementary particles have this property. Unlike Earth, however, they exist in the subatomic world and are governed by the laws of quantum mechanics. Therefore, these particles cannot spin in any arbitrary way, but only at certain specific rates. These rates can be $^1/_2$, 1, $^3/_2$, 2,... times a basic unit of rotation. Like protons and neutrons, electrons have spin $^1/_2$.

Particles with half-integer spin are called fermions, for the Italian American physicist Enrico Fermi, who investigated their properties in the first half of the 20th century. Fermions have one important property that will help explain both the way that electrons are arranged in their orbits and the way that protons and neutrons are arranged inside the nucleus. They are subject to the Pauli exclusion principle (named for the Austrian physicist Wolfgang Pauli), which states that no two fermions can occupy the same state—for example, the two electrons in a helium atom must have different spin directions if they occupy the same orbit.

Because a spinning electron can be thought of as a moving electric charge, electrons can be thought of as tiny electromagnets. This means that, like any other magnet, an electron will respond to the presence of a magnetic field by twisting. (Think of a compass needle pointing north under the influence of Earth's magnetic field.) This fact is usually expressed by saying that electrons have a magnetic moment. In physics, magnetic moment relates the strength of a magnetic field to the torque experienced by a magnetic object. Because of their intrinsic spin, electrons have a magnetic moment given by -9.28×10^{-24} joule per tesla.

Orbits and Energy Levels

Unlike planets orbiting the Sun, electrons cannot be at any arbitrary distance from the nucleus; they can exist only in certain specific locations called allowed orbits. This property, first explained by Danish physicist Niels Bohr in 1913, is another result of quantum mechanics—specifically, the requirement that the angular momentum of an electron in orbit, like everything else in the quantum world, come in discrete bundles called quanta.

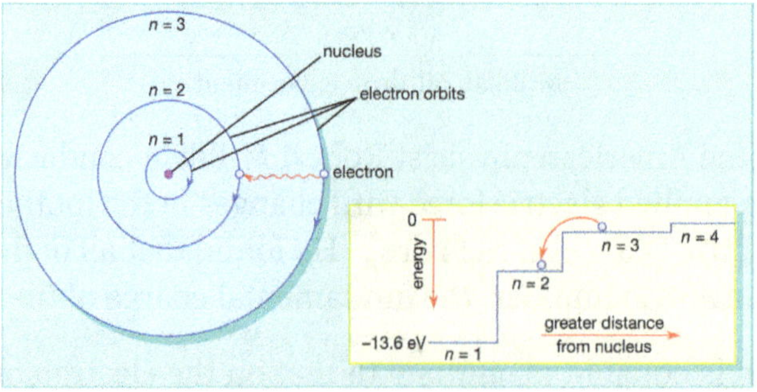

The Bohr atom.

The electron travels in circular orbits around the nucleus. The orbits have quantized sizes and energies. Energy is emitted from the atom when the electron jumps from one orbit to another closer to the nucleus. Shown here is the first Balmer transition, in which an electron jumps from orbit $n = 3$ to orbit $n = 2$, producing a photon of red light with an energy of 1.89 eV and a wavelength of 656 nanometres.

In the Bohr atom electrons can be found only in allowed orbits, and these allowed orbits are at different energies. The orbits are analogous to a set of stairs in which the gravitational potential energy is different for each step and in which a ball can be found on any step but never in between.

The laws of quantum mechanics describe the process by which electrons can move from one allowed orbit, or energy level, to another. As with many processes in the quantum world, this process is impossible to visualize. An electron disappears from the orbit in which it is located and reappears in its new location without ever appearing any place in between. This process is called a quantum leap or quantum jump, and it has no analog in the macroscopic world.

Because different orbits have different energies, whenever a quantum leap occurs, the energy possessed by the electron will be different after the jump. For example, if an electron jumps from a higher to a lower energy level, the lost energy will have to go somewhere and in fact will be emitted by the atom in a bundle of electromagnetic radiation. This bundle is known as a photon, and this emission of photons with a change of energy levels is the process by which atoms emit light.

In the same way, if energy is added to an atom, an electron can use that energy to make a quantum leap from a lower to a higher orbit. This energy can be supplied in many ways. One common way is for the atom to absorb a photon of just the right frequency. For example, when white light is shone on an atom, it selectively absorbs those frequencies corresponding to the energy differences between allowed orbits.

Each element has a unique set of energy levels, and so the frequencies at which it absorbs and emits light act as a kind of fingerprint, identifying the particular element. This property of atoms has given rise to spectroscopy, a science devoted to identifying atoms and molecules by the kind of radiation they emit or absorb.

This picture of the atom, with electrons moving up and down between allowed orbits, accompanied by the absorption or emission of energy, contains the essential features of the Bohr atomic model, for which Bohr received the Nobel Prize for Physics in 1922. His basic model does not work well in explaining the details of the structure of atoms more complicated than hydrogen, however. This requires the introduction of quantum mechanics. In quantum mechanics each orbiting electron is represented by a mathematical expression known as a wave function—something like a vibrating guitar string laid out along the path of the electron's orbit. These waveforms are called orbitals.

Electron Shells

In the quantum mechanical version of the Bohr atomic model, each of the allowed electron orbits is assigned a quantum number n that runs from 1 (for the orbit closest to the nucleus) to infinity (for orbits very far from the nucleus). All of the orbitals that have the same value of n make up a shell. Inside each shell there may be subshells corresponding to different rates of rotation and orientation of orbitals and the spin directions of the electrons. In general, the farther away from the nucleus a shell is the more subshells it will have.

This arrangement of possible orbitals explains a great deal about the chemical properties of different atoms. The easiest way to see this is to imagine building up complex atoms by starting with hydrogen and adding one proton and one electron (along with the appropriate number of neutrons) at a time. In hydrogen the lowest-energy orbit—called the ground state—corresponds to the electron located in the shell closest to the nucleus. There are two possible states for an electron in this shell, corresponding to a clockwise spin and a counterclockwise spin (or, in the jargon of physicists, spin up and spin down).

The next most-complex atom is helium, which has two protons in its nucleus and two orbiting electrons. These electrons fill the two available states in the lowest shell, producing what is called a filled shell. The next atom is lithium, with three electrons. Because the closest shell is filled, the third electron goes into the next higher shell. This shell has spaces for eight electrons, so that it takes an atom with 10 electrons (neon) to fill the first two levels. The next atom after neon, sodium, has 11 electrons, so that one electron goes into the next highest shell.

In the progression thus far, three atoms—hydrogen, lithium, and sodium—have one electron in the outermost shell. As stated above, it is these outermost electrons that determine the chemical properties of an atom. Therefore, these three elements should have similar properties, as indeed they do. For this reason, they appear in the same column of the periodic table of the elements, and the same principle determines the position of every element in that table. The outermost shell of electrons—called the valence shell—determines the chemical behaviour of an atom, and the number of electrons in this shell depends on how many are left over after all the interior shells are filled.

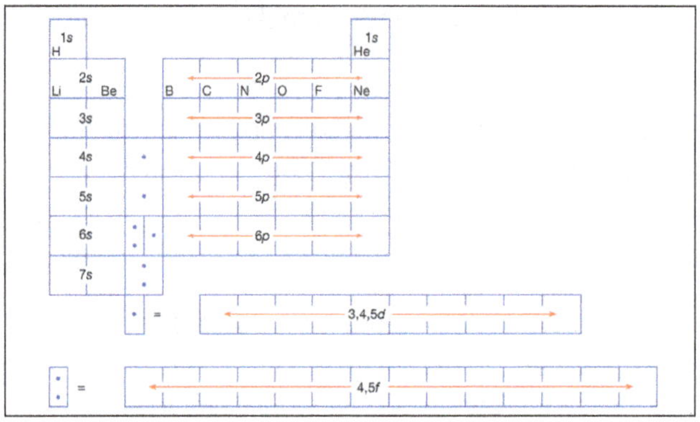

Periodic table of the elements showing the valence shells.

Atomic Bonds

Once the way atoms are put together is understood, the question of how they interact with each other can be addressed—in particular, how they form bonds to create molecules and macroscopic materials. There are three basic ways that the outer electrons of atoms can form bonds:

1. Electrons can be transferred from one atom to another.
2. Electrons can be shared between neighbouring atoms.
3. Electrons can be shared with all atoms in a material.

The first way gives rise to what is called an ionic bond. Consider as an example an atom of sodium, which has one electron in its outermost orbit, coming near an atom of chlorine, which has seven.

Because it takes eight electrons to fill the outermost shell of these atoms, the chlorine atom can be thought of as missing one electron. The sodium atom donates its single valence electron to fill the hole in the chlorine shell, forming a sodium chloride system at a lower total energy level.

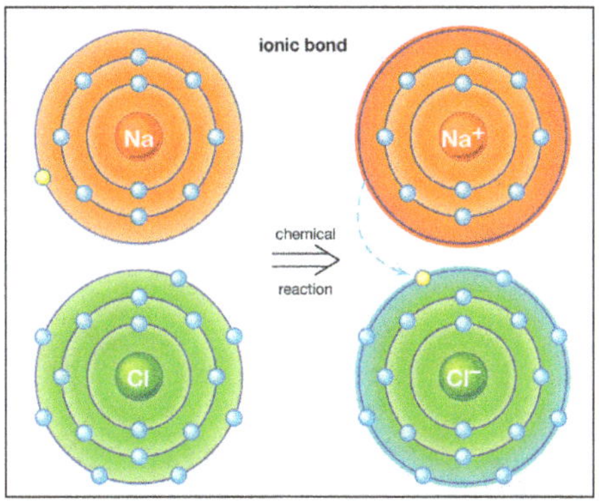

Ionic bond: sodium chloride or table salt.

The image above shows the Ionic bonding in sodium chloride. An atom of sodium (Na) donates one of its electrons to an atom of chlorine (Cl) in a chemical reaction, and the resulting positive ion (Na^+) and negative ion (Cl^-) form a stable ionic compound (sodium chloride; common table salt) based on this ionic bond.

An atom that has more or fewer electrons in orbit than protons in its nucleus is called an ion. Once the electron from its valence shell has been transferred, the sodium atom will be missing an electron; it therefore will have a positive charge and become a sodium ion. Simultaneously, the chlorine atom, having gained an extra electron, will take on a negative charge and become a chlorine ion. The electrical force between these two oppositely charged ions is attractive and locks them together. The resulting sodium chloride compound is a cubic crystal, commonly known as ordinary table salt.

The second bonding strategy listed is described by quantum mechanics. When two atoms come near each other, they can share a pair of outermost electrons (think of the atoms as tossing the electrons back and forth between them) to form a covalent bond. Covalent bonds are particularly common in organic materials, where molecules often contain long chains of carbon atoms (which have four electrons in their valence shells).

Finally, in some materials each atom gives up an outer electron that then floats freely—in essence, the electron is shared by all of the atoms within the material. The electrons form a kind of sea in which the positive ions float like marbles in molasses. This is called the metallic bond and, as the name implies, it is what holds metals together.

There are also ways for atoms and molecules to bond without actually exchanging or sharing electrons. In many molecules the internal forces are such that the electrons tend to cluster at one end of the molecule, leaving the other end with a positive charge. Overall, the molecule has no net electric charge—it is just that the positive and negative charges are found at different places. For example, in water (H_2O) the electrons tend to spend most of their time near the oxygen atom, leaving the region of the hydrogen atoms with a positive charge. Molecules whose charges are arranged in this way are called polar molecules. An atom or ion approaching a polar molecule from its negative side, for example, will experience a stronger negative electric force than the more-distant positive electric force. This is why many substances dissolve in water: the polar water molecule can pull ions out of materials by exerting electric forces. A special case of polar forces occurs in what is called the hydrogen bond. In many situations, when hydrogen forms a covalent bond with another atom, electrons move toward that atom, and the hydrogen acquires a slight positive charge. The hydrogen, in turn, attracts another atom, thereby forming a kind of bridge between the two. Many important molecules, including DNA, depend on hydrogen bonds for their structure.

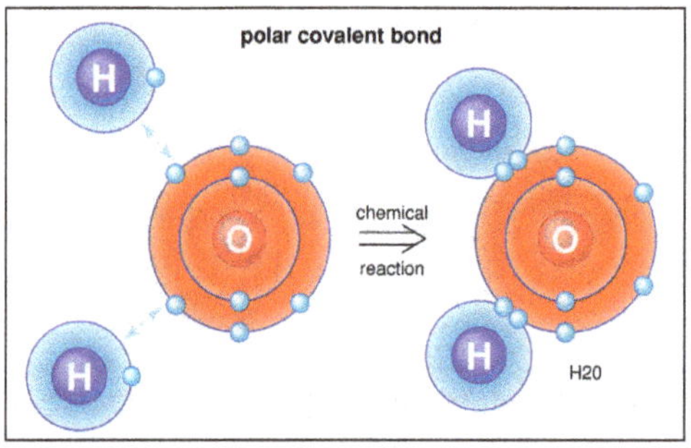

Polar covalent bond.

In polar covalent bonds, such as that between hydrogen and oxygen atoms, the electrons are not transferred from one atom to the other as they are in an ionic bond. Instead, some outer electrons merely spend more time in the vicinity of the other atom. The effect of this orbital distortion is to induce regional net charges that hold the atoms together, such as in water molecules.

Finally, there is a way for a weak bond to form between two electrically neutral atoms. Dutch physicist Johannes van der Waals first theorized a mechanism for such a bond in 1873, and it is now known as van der Waals forces. When two atoms approach each other, their electron clouds exert repulsive forces on each other, so that the atoms become polarized. In such situations, it is possible that the electrical attraction between the nucleus of one atom and the electrons of the other will overcome the repulsive forces between the electrons, and a weak bond will form. One example of this force can be seen in ordinary graphite pencil lead. In this material, carbon atoms are held together in sheets by strong covalent bonds, but the sheets are held together only by van der Waals forces. When a pencil is drawn across paper, the van der Waals forces break, and sheets of carbon slough off. This is what creates the dark pencil streak.

Conductors and Insulators

The way that atoms bond together affects the electrical properties of the materials they form. For example, in materials held together by the metallic bond, electrons float loosely between the metal ions. These electrons will be free to move if an electrical force is applied. For example, if a copper wire is attached across the poles of a battery, the electrons will flow inside the wire. Thus, an electric current flows, and the copper is said to be a conductor.

The flow of electrons inside a conductor is not quite so simple, though. A free electron will be accelerated for a while but will then collide with an ion. In the collision process, some of the energy acquired by the electron will be transferred to the ion. As a result, the ion will move faster, and an observer will notice the wire's temperature rise. This conversion of electrical energy from the motion of the electrons to heat energy is called electrical resistance. In a material of high resistance, the wire heats up quickly as electric current flow. In a material of low resistance, such as copper wire, most of the energy remains with the moving electrons, so the material is good at moving electrical energy from one point to another. Its excellent conducting property, together with its relatively low cost, is why copper is commonly used in electrical wiring.

The exact opposite situation obtains in materials, such as plastics and ceramics, in which the electrons are all locked into ionic or covalent bonds. When these kinds of materials are placed between the poles of a battery, no current flows—there are simply no electrons free to move. Such materials are called insulators.

Magnetic Properties

The magnetic properties of materials are also related to the behaviour of electrons in atoms. An electron in orbit can be thought of as a miniature loop of electric current. According to the laws of electromagnetism, such a loop will create a magnetic field. Each electron in orbit around a nucleus produces its own magnetic field, and the sum of these fields, together with the intrinsic fields of the electrons and the nucleus, determines the magnetic field of the atom. Unless all of these fields cancel out, the atom can be thought of as a tiny magnet.

In most materials these atomic magnets point in random directions, so that the material itself is not magnetic. In some cases—for instance, when randomly oriented atomic magnets are placed in a strong external magnetic field—they line up, strengthening the external field in the process. This phenomenon is known as paramagnetism. In a few metals, such as iron, the interatomic forces are such that the atomic magnets line up over regions a few thousand atoms across. These regions are called domains. In normal iron the domains are oriented randomly, so the material is not magnetic. If iron is put in a strong magnetic field, however, the domains will line up, and they will stay lined up even after the external field is removed. As a result, the piece of iron will acquire a strong magnetic field. This phenomenon is known as ferromagnetism. Permanent magnets are made in this way.

The Nucleus

Nuclear Forces

The primary constituents of the nucleus are the proton and the neutron, which have approximately equal mass and are much more massive than the electron. For reference, the accepted mass of the proton is $1.672621777 \times 10^{-24}$ gram while that of the neutron is $1.674927351 \times 10^{-24}$ gram. The charge on the proton is equal in magnitude to that on the electron but is opposite in sign, while the neutron has no electrical charge. Both particles have spin $^1/_2$ and are therefore fermions and subject to the Pauli exclusion principle. Both also have intrinsic magnetic fields. The magnetic moment of the proton is $1.410606743 \times 10^{-26}$ joule per tesla, while that of the neutron is $-0.96623647 \times 10^{-26}$ joule per tesla.

It would be incorrect to picture the nucleus as just a collection of protons and neutrons, analogous to a bag of marbles. In fact, much of the effort in physics research during the second half of the 20th century was devoted to studying the various kinds of particles that live out their fleeting lives inside the nucleus. A more-accurate picture of the nucleus would be of a seething cauldron where hundreds of different kinds of particles swarm around the protons and neutrons. It is now believed that these so-called elementary particles are made of still more-elementary objects, which have been given the name of quarks. Modern theories suggest that even the quarks may be made of still more-fundamental entities called strings.

The forces that operate inside the nucleus are a mixture of those familiar from everyday life and those that operate only inside the atom. Two protons, for example, will repel each other because of their identical electrical force but will be attracted to each other by gravitation. Especially at the scale of elementary particles, the gravitational force is many orders of magnitude weaker than other fundamental forces, so it is customarily ignored when talking about the nucleus. Nevertheless, because the nucleus stays together in spite of the repulsive electrical force between protons, there must exist a counterforce—which physicists have named the strong force—operating at short range within the nucleus. The strong force has been a major concern in physics research since its existence was first postulated in the 1930s.

One more force—the weak force—operates inside the nucleus. The weak force is responsible for some of the radioactive decays of nuclei. The four fundamental forces—strong, electromagnetic, weak, and gravitational—are responsible for every process in the universe. One of the important strains in modern theoretical physics is the idea that, although they seem very different, they are different aspects of a single underlying force.

Nuclear Shell Model

Many models describe the way protons and neutrons are arranged inside a nucleus. One of the most successful and simple to understand is the shell model. In this model the protons and neutrons occupy separate systems of shells, analogous to the shells in which electrons are found outside the nucleus. From light to heavy nuclei, the proton and neutron shells are filled (separately) in much the same way as electron shells are filled in an atom.

Like the Bohr atomic model, the nucleus has energy levels that correspond to processes in which protons and neutrons make quantum leaps up and down between their allowed orbits. Because energies in the nucleus are so much greater than those associated with electrons, however, the photons emitted or absorbed in these reactions tend to be in the X-ray or gamma ray portions of the electromagnetic spectrum, rather than the visible light portion.

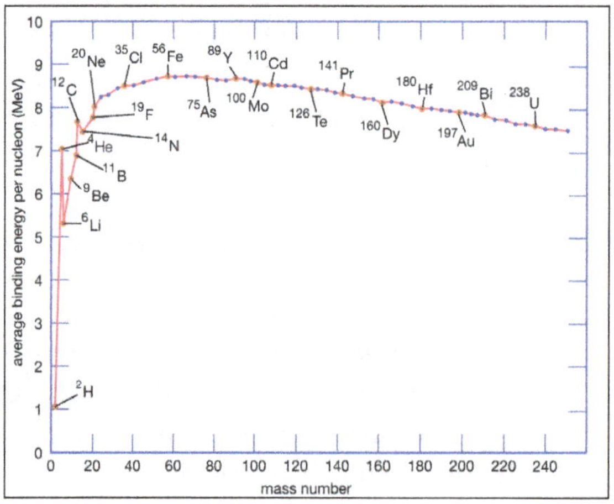

Nuclear binding energies, shown as a function of atomic mass number.

When a nucleus forms from protons and neutrons, an interesting regularity can be seen: the mass of the nucleus is slightly less than the sum of the masses of the constituent protons and neutrons. This consistent discrepancy is not large—typically only a fraction of a percent—but it is significant. By Albert Einstein's principles of relativity, this small mass deficit can be converted into energy via the equation $E = mc^2$. Thus, in order to break a nucleus into its constituent protons and neutrons, energy must be supplied to make up this mass deficit. The energy corresponding to the mass deficit is called the binding energy of the nucleus, and as the name suggests, it represents the energy required to tie the nucleus together. The binding energy varies across the periodic table and is at a maximum for iron, which is thus the most stable element.

Radioactive Decay

The nuclei of most everyday atoms are stable—that is, they do not change over time. This statement is somewhat misleading, however, because nuclei that are not stable generally do not last long and hence tend not to be part of everyday experience. In fact, most of the known isotopes of nuclei are not stable; instead, they go through a process called radioactive decay, which often changes the identity of the original atom.

In radioactive decay a nucleus will remain unchanged for some unpredictable period and then emit a high-speed particle or photon, after which a different nucleus will have replaced the original. Each unstable isotope decays at a different rate; that is, each has a different probability of decaying within a given period of time. A collection of identical unstable nuclei do not all decay at once. Instead, like popcorn popping in a pan, they will decay individually over a period of time. The time that it takes for half of the original sample to decay is called the half-life of the isotope. Half-lives of known isotopes range from microseconds to billions of years. Uranium-238 (^{238}U) has a half-life of about 4.5 billion years, which is approximately the time that has elapsed since the formation of the solar system. Thus, Earth has about half of the ^{238}U that it had when it was formed.

There are three different types of radioactive decay. In the late 19th century, when radiation was still mysterious, these forms of decay were denoted alpha, beta, and gamma. In alpha decay a nucleus ejects two protons and two neutrons, all locked together in what is called an alpha particle (later discovered to be identical to the nucleus of a normal helium atom). The daughter, or decayed, nucleus will have two fewer protons and two fewer neutrons than the original and hence will be the nucleus of a different chemical element. Once the electrons have rearranged themselves (and the two excess electrons have wandered off), the atom will, in fact, have changed identity.

In beta decay one of the neutrons in the nucleus turns into a proton, a fast-moving electron, and a particle called a neutrino. This emission of fast electrons is called beta radiation. The daughter nucleus has one fewer neutron and one more proton than the original and hence, again is a different chemical element.

In gamma decay a proton or neutron makes a quantum leap from a higher to a lower orbit, emitting a high-energy photon in the process. In this case the chemical identity of the daughter nucleus is the same as the original.

When a radioactive nucleus decays, it often happens that the daughter nucleus is radioactive as well. This daughter will decay in turn, and the daughter nucleus of that decay may be radioactive as well. Thus, a collection of identical atoms may, over time, be turned into a mixture of many kinds of atoms because of successive decays. Such decays will continue until stable daughter nuclei are produced. This process, called a decay chain, operates everywhere in nature. For example, uranium-238 decays with a half-life of 4.5 billion years into thorium-234, which decays in 24 days into protactinium-234, which also decays. This process continues until it gets to lead-206, which is stable. Dangerous elements such as radium and radon are continually produced in Earth's crust as intermediary steps in decay chains.

Nuclear Energy

It is almost impossible to have lived at any time since the mid-20th century and not be aware that energy can be derived from the atomic nucleus. The basic physical principle behind this fact is that the total mass present after a nuclear reaction is less than before the reaction. This difference in mass, via the equation $E = mc^2$, is converted into what is called nuclear energy.

There are two types of nuclear processes that can produce energy—nuclear fission and nuclear fusion. In fission a heavy nucleus (such as uranium) is split into a collection of lighter nuclei and fast-moving particles. The energy at the end typically appears in the kinetic energy of the final particles. Nuclear fission is used in nuclear reactors to produce commercial electricity. It depends on the fact that a particular isotope of uranium (^{235}U) behaves in a particular way when it is hit by a neutron. The nucleus breaks apart and emits several particles. Included in the debris of the fission are two or three more free neutrons that can produce fission in other nuclei in a chain reaction. This chain reaction can be controlled and used to heat water into steam, which can then be used to turn turbines in an electrical generator.

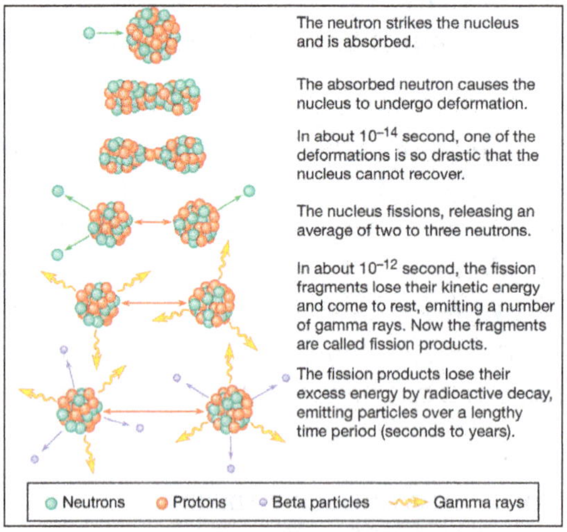
The neutron strikes the nucleus and is absorbed.
The absorbed neutron causes the nucleus to undergo deformation.
In about 10^{-14} second, one of the deformations is so drastic that the nucleus cannot recover.
The nucleus fissions, releasing an average of two to three neutrons.
In about 10^{-12} second, the fission fragments lose their kinetic energy and come to rest, emitting a number of gamma rays. Now the fragments are called fission products.
The fission products lose their excess energy by radioactive decay, emitting particles over a lengthy time period (seconds to years).
Neutrons
Protons
Beta particles
Gamma rays

Sequence of events in the fission of a uranium nucleus by a neutron.

Fusion refers to a process in which two or more light nuclei come together to form a heavier nucleus. The most common fusion process in nature is one in which four protons come together to form a helium nucleus (two protons and two neutrons) and some other particles. This is the process by which energy is generated in stars. Scientists have not yet learned to produce a controllable, commercially useful nuclear fusion on Earth, which remains a goal for the future.

Studies of the Properties of Atoms

Size of Atoms

The first modern estimates of the size of atoms and the numbers of atoms in a given volume were made by German chemist Joseph Loschmidt in 1865. Loschmidt used the results of kinetic theory and some rough estimates to do his calculation. The size of the atoms and the distance between them in the gaseous state are related both to the contraction of gas upon liquefaction and to the mean free path traveled by molecules in a gas. The mean free path, in turn, can be found from the thermal conductivity and diffusion rates in the gas. Loschmidt calculated the size of the atom and the spacing between atoms by finding a solution common to these relationships. His result for Avogadro's number is remarkably close to the present accepted value of about 6.022×10^{23}. The precise definition of Avogadro's number is the number of atoms in 12 grams of the carbon isotope C-12. Loschmidt's result for the diameter of an atom was approximately 10^{-8} cm.

Much later, in 1908, French physicist Jean Perrin used Brownian motion to determine Avogadro's number. Brownian motion, first observed in 1827 by Scottish botanist Robert Brown, is the continuous movement of tiny particles suspended in water. Their movement is caused by the thermal motion of water molecules bumping into the particles. Perrin's argument for determining Avogadro's number makes an analogy between particles in the liquid and molecules in the atmosphere. The thinning of air at high altitudes depends on the balance between the gravitational force pulling the molecules down and their thermal motion forcing them up. The relationship between the weight of the particles and the height of the atmosphere would be the same for Brownian particles suspended in water. Perrin counted particles of gum mastic at different heights in his water sample and inferred the mass of atoms from the rate of decrease. He then divided the result into the molar weight of atoms to determine Avogadro's number. After Perrin, few scientists could disbelieve the existence of atoms.

Electric Properties of Atoms

While atomic theory was set back by the failure of scientists to accept simple physical ideas like the diatomic molecule and the kinetic theory of gases, it was also delayed by the preoccupation of physicists with mechanics for almost 200 years, from Newton to the 20th century. Nevertheless, several 19th-century investigators, working in the relatively ignored fields of electricity, magnetism, and optics, provided important clues about the interior of the atom. The studies in electrodynamics made by English physicist Michael Faraday and those of Maxwell indicated for the first time that something existed apart from palpable matter, and data obtained by Gustav Robert Kirchhoff of Germany about elemental spectral lines raised questions that would be answered only in the 20th century by quantum mechanics.

Until Faraday's electrolysis experiments, scientists had no conception of the nature of the forces binding atoms together in a molecule. Faraday concluded that electrical forces existed inside the molecule after he had produced an electric current and a chemical reaction in a solution with the electrodes of a voltaic cell. No matter what solution or electrode material he used, a fixed quantity of current sent through an electrolyte always caused a specific amount of material to form on an electrode of the electrolytic cell. Faraday concluded that each ion of a given chemical compound has exactly the same charge. Later he discovered that the ionic charges are integral multiples of a single unit of charge, never fractions.

On the practical level, Faraday did for charge what Dalton had done for the chemical combination of atomic masses. That is to say, Faraday demonstrated that it takes a definite amount of charge to convert an ion of an element into an atom of the element and that the amount of charge depends on the element used. The unit of charge that releases one gram-equivalent weight of a simple ion is called the faraday in his honour. For example, one faraday of charge passing through water releases one gram of hydrogen and eight grams of oxygen. In this manner, Faraday gave scientists a rather precise value for the ratios of the masses of atoms to the electric charges of ions. The ratio of the mass of the hydrogen atom to the charge of the electron was found to be 1.035×10^{-8} kilogram per coulomb. Faraday did not know the size of his electrolytic unit of charge in units such as coulombs any more than Dalton knew the magnitude of his unit of atomic weight in grams. Nevertheless, scientists could determine the ratio of these units easily.

More significantly, Faraday's work was the first to imply the electrical nature of matter and the existence of subatomic particles and a fundamental unit of charge. Faraday did not, however, conclude that atoms cause electricity.

Light and Spectral Lines

In 1865, Maxwell unified the laws of electricity and magnetism in his publication "A Dynamical Theory of the Electromagnetic Field." He concluded that light is an electromagnetic wave. His theory was confirmed by German physicist Heinrich Hertz, who produced radio waves with sparks in 1887. With light understood as an electromagnetic wave, Maxwell's theory could be applied to the emission of light from atoms. The theory failed, however, to describe spectral lines and the fact that atoms do not lose all their energy when they radiate light. The problem was not with Maxwell's theory of light itself but rather with its description of the oscillating electron currents generating light. Only quantum mechanics could explain this behaviour.

By far the richest clues about the structure of the atom came from spectral line series. Mounting a particularly fine prism on a telescope, German physicist and optician Joseph von Fraunhofer had discovered between 1814 and 1824 hundreds of dark lines in the spectrum of the Sun. He labeled the most prominent of these lines with the letters A through G. Together they are now called Fraunhofer lines. A generation later Kirchhoff heated different elements to incandescence in order to study the different coloured vapours emitted. Observing the vapours through a spectroscope, he discovered that each element has a unique and characteristic pattern of spectral lines. Each element produces the same set of identifying lines, even when it is combined chemically with other elements. In 1859, Kirchhoff and German chemist Robert Wilhelm Bunsen discovered two new elements—cesium and rubidium—by first observing their spectral lines.

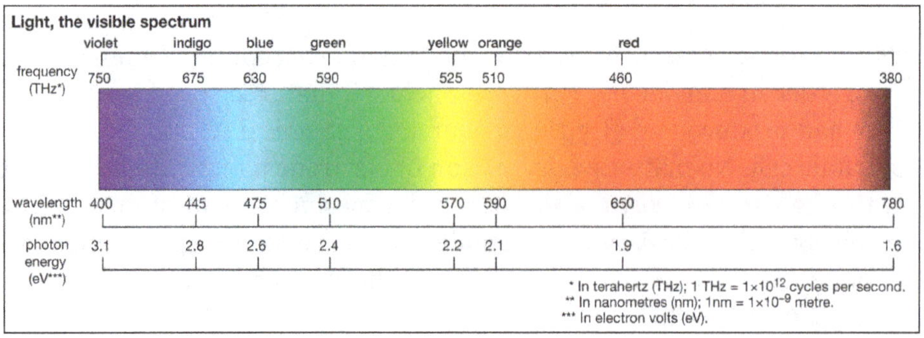

The figure shows the visible solar spectrum, ranging from the shortest visible wavelengths (violet light, at 400 nm) to the longest (red light, at 700 nm). Shown in the diagram are prominent Fraunhofer lines, representing wavelengths at which light is absorbed by elements present in the atmosphere of the Sun.

Johann Jakob Balmer, a Swiss secondary-school teacher with a penchant for numerology, studied hydrogen's spectral lines and found a constant relationship between the wavelengths of the element's four visible lines. In 1885, he published a generalized mathematical formula for all the lines of hydrogen. Swedish physicist Johannes Rydberg extended Balmer's work in 1890 and found a general rule applicable to many elements. Soon more series were discovered elsewhere in the spectrum of hydrogen and in the spectra of other elements as well. Stated in terms of the frequency of the light rather than its wavelength, the formula may be expressed:

$$v = R(1/n^2 - 1/m^2).$$

Here v is the frequency of the light, n and m are integers, and R is the Rydberg constant. In the Balmer lines m is equal to 2 and n takes on the values 3, 4, 5, and 6.

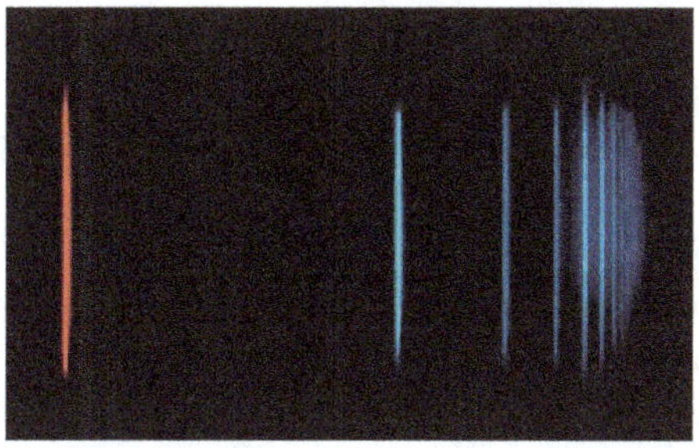

The Balmer series of hydrogen as seen by a low-resolution spectrometer.

Discovery of Electrons

During the 1880s and '90s scientists searched cathode rays for the carrier of the electrical properties in matter. Their work culminated in the discovery by English physicist J.J. Thomson of the electron in 1897. The existence of the electron showed that the 2,000-year-old conception of the atom as a homogeneous particle was wrong and that in fact the atom has a complex structure.

Cathode-ray studies began in 1854 when Heinrich Geissler, a glassblower and technical assistant to German physicist Julius Plücker, improved the vacuum tube. Plücker discovered cathode rays in 1858 by sealing two electrodes inside the tube, evacuating the air, and forcing electric current between the electrodes. He found a green glow on the wall of his glass tube and attributed it to rays emanating from the cathode. In 1869, with better vacuums, Plücker's pupil Johann W. Hittorf saw a shadow cast by an object placed in front of the cathode. The shadow proved that the cathode rays originated from the cathode. English physicist and chemist William Crookes investigated cathode rays in 1879 and found that they were bent by a magnetic field; the direction of deflection suggested that they were negatively charged particles. As the luminescence did not depend on what gas had been in the vacuum or what metal the electrodes were made of, he surmised that the rays were a property of the electric current itself. As a result of Crookes's work, cathode rays were widely studied, and the tubes came to be called Crookes tubes.

Although Crookes believed that the particles were electrified charged particles, his work did not settle the issue of whether cathode rays were particles or radiation similar to light. By the late 1880s, the controversy over the nature of cathode rays had divided the physics community into two camps. Most French and British physicists, influenced by Crookes, thought that cathode rays were electrically charged particles because they were affected by magnets. Most German physicists, on the other hand, believed that the rays were waves because they traveled in straight lines and were unaffected by gravity. A crucial test of the nature of the cathode rays was how they would be affected by electric fields. Heinrich Hertz, the aforementioned German physicist, reported that the cathode rays were not deflected when they passed between two oppositely charged plates in an 1892 experiment. In England, J.J. Thomson thought Hertz's vacuum might have been faulty and that residual gas might have reduced the effect of the electric field on the cathode rays.

Thomson repeated Hertz's experiment with a better vacuum in 1897. He directed the cathode rays between two parallel aluminum plates to the end of a tube where they were observed as luminescence on the glass. When the top aluminum plate was negative, the rays moved down; when the upper plate was positive, the rays moved up. The deflection was proportional to the difference in potential between the plates. With both magnetic and electric deflections observed, it was clear that cathode rays were negatively charged particles. Thomson's discovery established the particulate nature of electricity. Accordingly, he called his particles electrons.

From the magnitude of the electrical and magnetic deflections, Thomson could calculate the ratio of mass to charge for the electrons. This ratio was known for atoms from electrochemical studies. Measuring and comparing it with the number for an atom, he discovered that the mass of the electron was very small, merely 1/1,836 that of a hydrogen ion. When scientists realized that an electron was virtually 1,000 times lighter than the smallest atom, they understood how cathode rays could penetrate metal sheets and how electric current could flow through copper wires. In deriving the mass-to-charge ratio, Thomson had calculated the electron's velocity. It was $^1/_{10}$ the speed of light, thus amounting to roughly 30,000 km (18,000 miles) per second. Thomson emphasized that:

> " We have in the cathode rays matter in a new state, a state in which the subdivision of matter is carried very much further than in the ordinary gaseous state; a state in which all matter, that is, matter derived from different sources such as hydrogen, oxygen, etc., is of one and the same kind; this matter being the substance from which all the chemical elements are built up."

Thus, the electron was the first subatomic particle identified the smallest and the fastest bit of matter known at the time.

In 1909, American physicist Robert Andrews Millikan greatly improved a method employed by Thomson for measuring the electron charge directly. In Millikan's oil-drop experiment, he produced microscopic oil droplets and observed them falling in the space between two electrically charged plates. Some of the droplets became charged and could be suspended by a delicate adjustment of the electric field. Millikan knew the weight of the droplets from their rate of fall when the electric field was turned off. From the balance of the gravitational and electrical forces, he could determine the charge on the droplets. All the measured charges were integral multiples of a quantity that in contemporary units is 1.602×10^{-19} coulomb. Millikan's electron-charge experiment was the first to detect and measure the effect of an individual subatomic particle. Besides confirming the particulate nature of electricity, his experiment also supported previous determinations of Avogadro's number. Avogadro's number times the unit of charge gives Faraday's constant, the amount of charge required to electrolyze one mole of a chemical ion.

Identification of Positive Ions

In addition to electrons, positively charged particles also emanate from the anode in an energized Crookes tube. German physicist Wilhelm Wien analyzed these positive rays in 1898 and found that the particles have a mass-to-charge ratio more than 1,000 times larger than that of the electron. Because the ratio of the particles is also comparable to the mass-to-charge ratio of the residual atoms in the discharge tubes, scientists suspected that the rays were actually ions from the gases in the tube.

In 1913, Thomson refined Wien's apparatus to separate different ions and measure their mass-to-charge ratio on photographic plates. He sorted out the many ions in various charge states produced in a discharge tube. When he conducted his atomic mass experiments with neon gas, he found that a beam of neon atoms subjected to electric and magnetic forces split into two parabolas instead of one on a photographic plate. Chemists had assumed the atomic weight of neon was 20.2, but the traces on Thomson's photographic plate suggested atomic weights of 20.0 and 22.0, with the former parabola much stronger than the latter. He concluded that neon consisted of two stable isotopes: primarily neon-20, with a small percentage of neon-22. Eventually a third isotope, neon-21, was discovered in very small quantities. It is now known that 1,000 neon atoms will contain an average of 909 atoms of neon-20, 88 of neon-22, and 3 of neon-21. Dalton's assumptions that all atoms of an element have an identical mass and that the atomic weight of an element is its mass were thus disproved. Today the atomic weight of an element is recognized as the weighted average of the masses of its isotopes.

Francis William Aston, an English physicist, improved Thomson's technique when he developed the mass spectrograph in 1919. This device spread out the beam of positive ions into a "mass spectrum" of lines similar to the way light is separated into a spectrum. Aston analyzed about 50 elements over the next six years and discovered that most have isotopes.

Discovery of Radioactivity

Like Thomson's discovery of the electron, the discovery of radioactivity in uranium by French physicist Henri Becquerel in 1896 forced scientists to radically change their ideas about atomic structure. Radioactivity demonstrated that the atom was neither indivisible nor immutable. Instead of serving merely as an inert matrix for electrons, the atom could change form and emit an enormous amount of energy. Furthermore, radioactivity itself became an important tool for revealing the interior of the atom.

German physicist Wilhelm Conrad Röntgen had discovered X-rays in 1895, and Becquerel thought they might be related to fluorescence and phosphorescence, processes in which substances absorb and emit energy as light. In the course of his investigations, Becquerel stored some photographic plates and uranium salts in a desk drawer. Expecting to find the plates only lightly fogged, he developed them and was surprised to find sharp images of the salts. He then began experiments that showed that uranium salts emit a penetrating radiation independent of external influences. Becquerel also demonstrated that the radiation could discharge electrified bodies. In this case discharge means the removal of electric charge, and it is now understood that the radiation, by ionizing molecules of air, allows the air to conduct an electric current. Early studies of radioactivity relied on measuring ionization power or on observing the effects of radiation on photographic plates.

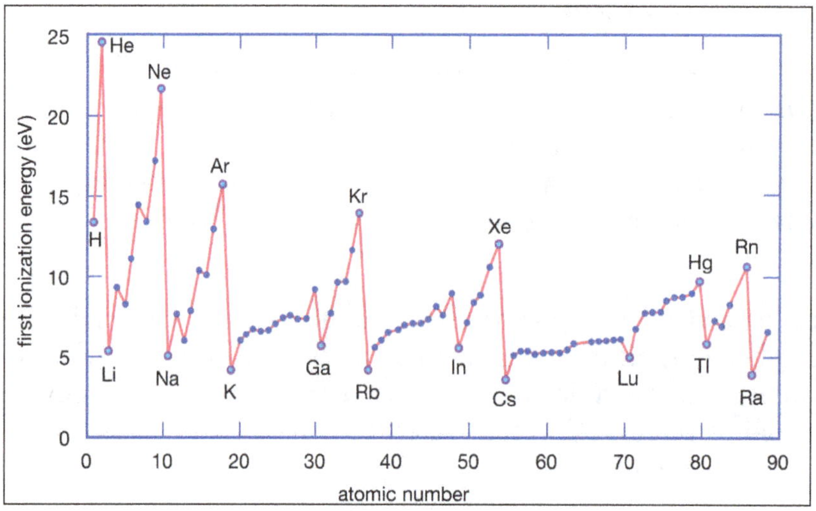

First ionization energies of the elements.

In 1898, French physicists Pierre and Marie Curie discovered the strongly radioactive elements polonium and radium, which occur naturally in uranium minerals. Marie coined the term radioactivity for the spontaneous emission of ionizing, penetrating rays by certain atoms.

Experiments conducted by British physicist Ernest Rutherford in 1899 showed that radioactive substances emit more than one kind of radiation. It was determined that part of the radiation is 100 times more penetrating than the rest and can pass through aluminum foil one-fiftieth of a millimetre thick. Rutherford named the less-penetrating emanations alpha rays and the more-powerful ones beta rays, after the first two letters of the Greek alphabet. Investigators who in 1899 found that beta rays were deflected by a magnetic field concluded that they are negatively charged particles similar to cathode rays. In 1903, Rutherford found that alpha rays were deflected slightly in the opposite direction, showing that they are massive, positively charged particles. Much later Rutherford proved that alpha rays are nuclei of helium atoms by collecting the rays in an evacuated tube and detecting the buildup of helium gas over several days.

A third kind of radiation was identified by French chemist Paul Villard in 1900. Designated as the gamma ray, it is not deflected by magnets and is much more penetrating than alpha particles. Gamma rays were later shown to be a form of electromagnetic radiation, similar to light or X-rays, but with much shorter wavelengths. Because of these shorter wavelengths, gamma rays have higher frequencies and are even more penetrating than X-rays.

In 1902, while studying the radioactivity of thorium, Rutherford and English chemist Frederick Soddy discovered that radioactivity was associated with changes inside the atom that transformed thorium into a different element. They found that thorium continually generates a chemically different substance that is intensely radioactive. The radioactivity eventually makes the new element disappear. Watching the process, Rutherford and Soddy formulated the exponential decay law, which states that a fixed fraction of the element will decay in each unit of time. For example, half of the thorium product decays in four days, half the remaining sample in the next four days, and so on.

Until the 20th century, physicists had studied subjects, such as mechanics, heat, and electromagnetism that they could understand by applying common sense or by extrapolating from everyday experiences. The discoveries of the electron and radioactivity, however, showed that classical Newtonian mechanics could not explain phenomena at atomic and subatomic levels. As the primacy of classical mechanics crumbled during the early 20th century, quantum mechanics was developed to replace it. Since then experiments and theories have led physicists into a world that is often extremely abstract and seemingly contradictory.

Proton

In physics, the proton is a subatomic particle with an electric charge of one positive fundamental unit. The proton is observed to be stable, although some theories predict that the proton may decay. The proton has a density of about 2.31×10^{17} kg m^{-3}.

Protons are spin-1/2 fermions and are composed of three quarks, making them baryons. The two up quarks and one down quark of the proton are also held together by the strong nuclear force, mediated by gluons. Protons may be transmuted into neutrons by inverse beta decay (that is, by capturing an electron); since neutrons are heavier than protons, this process does not occur spontaneously but only when energy is supplied. The proton's anti-matter equivalent is the antiproton, which has the same magnitude charge as the proton but the opposite sign.

Protons and neutrons are both nucleons, which may be bound by the nuclear force into atomic nuclei. The most common isotope of the hydrogen atom is a single proton. The nuclei of other atoms are composed of various numbers of protons and neutrons. The number of protons in the nucleus determines the chemical properties of the atom and which chemical element it is.

In chemistry and biochemistry, the proton is thought of as the hydrogen ion, denoted H^+. In this context, a proton donor is an acid and a proton acceptor a base.

The proton is the simplest composite particle. It contains three fermions, called quarks, that have shed all their color charge into a halo of gluons as dictated by the quantum probability wavefunction of the color interaction.

A gluon is similar to a photon of light, except that a gluon goes both backwards and forwards in time, it has a quantum 'color' charge at one end, and a quantum 'anticolor' at the other end. In the middle, the gluon is colorless. The gluon halo places all the color and anticolor at the surface of the proton, while the center of the proton is colorless. This colorless center is where the now essentially colorless quarks spend most of the time. The quarks are 'confined' to this colorless center but have plenty of space to move around in as size of the quarks compared to the extended gluon halo is as three dust particles are to New York City.

The surface of the proton, as far as the color interaction is concerned, can be compared to a color computer screen displaying white. On close inspection, however, it is composed of red, blue and green pixels. This is why the color surface of the proton with its pixels of gluon ends appears colorless. The energy of this relatively vast gluon halo is responsible for 99.9 percent of the mass-energy of the proton.

The quarks, while shedding their color charge, have no way to shed their electric charge. As the charged quarks are confined to the colorless central regions of the proton, the electric diameter of the proton—the region where all the electric charge is concentrated—is significantly smaller than the color-charge diameter of the proton.

There are two types of quarks in regular matter. The three quarks in a proton are two U-quarks each with +2/3 electric charge, and a D-quark with -1/3 charge. The composite proton has an overall charge of +1. In a neutron, the other composite particle found in atomic nuclei, there is one U and 2 Ds, and the neutron has zero overall charge. An aspect of the color interaction akin to polarization in light photons makes the neutron combination of quarks generates a few more gluons than the proton combination, and this gives the neutron slightly more mass-energy than the proton.

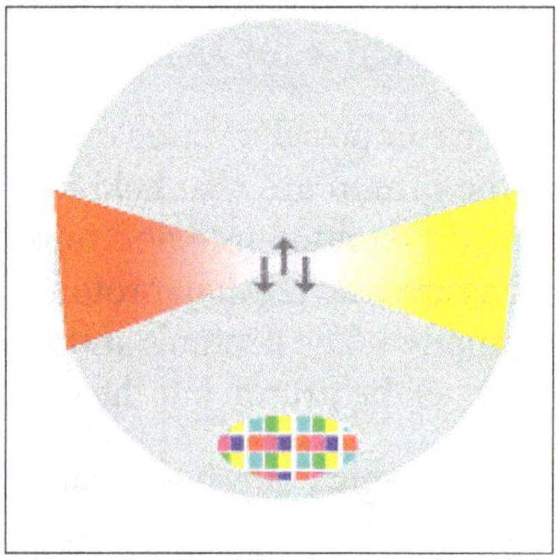

A representation of a proton.

This is a diagram of a proton, with one of the gluons magnified. This particular gluon—one of the eight possible combinations of color-anticolor, has quantum red going forward in time and quantum blue (as its anticolor complement, yellow) going backwards in time. As long as the (sand-grain sized) quarks stay in the colorless center, the experience what is called 'asymptotic freedom' and are free of the color influence on them. In this state, they align themselves according to the quantum waveform of their electromagnetic interaction as they freely couple photons.

For the quark to leave the colorless center and pick up color from the periphery, however, takes a lot of energy. So much energy, in fact, that (virtual particle) pairs of quarks and antiquarks become real when a quark gets kicked out of the center (perhaps by a very energetic electron) and new combinations of quarks swathed in gluons appear, such as (pions) and other such combinations. At no time in any such process is there a "bare quark" which can be observed. Quarks are confined by what is called "infrared slavery" (they cannot escape their low energy colorless state) to always be observed in composite, colorless combinations.

Antiproton

The antiproton is the antiparticle of the proton. It was discovered in 1955 by Emilio Segre and Owen Chamberlain, for which they were awarded the 1959 Nobel Prize in Physics.

CPT-symmetry puts strong constraints on the relative properties of particles and antiparticles and, therefore, is open to stringent tests. For example, the charges of the proton and antiproton must sum to exactly zero. This equality has been tested to one part in 10^8. The equality of their masses is also tested to better than one part in 10^8. By holding antiprotons in a Penning trap, the equality of the charge to mass ratio of the proton and the antiproton has been tested to 1 part in 9×10^{11}. The magnetic moment of the antiproton has been measured with error of 8×10^{-3} nuclear Bohr magnetons, and is found to be equal and opposite to that of the proton.

Neutron

The neutron is a subatomic particle that is an important constituent of all atomic nuclei with one exception, that of hydrogen. It is important for the stability of nuclei that have more than one proton. As such it is one of the basic building blocks of the physical universe. Though it has no electrical charge it is composed of charged particles (quarks), whose combined charges cancel out. The neutron thus refects a relational balance of charge rather than an absence of it. It has a mass of 939.573 MeV/c^2 (1.6749 $\times 10^{-27}$ kg, slightly more than a proton). Its spin is ½. Its antiparticle is called the antineutron. The neutron and proton are both nucleons.

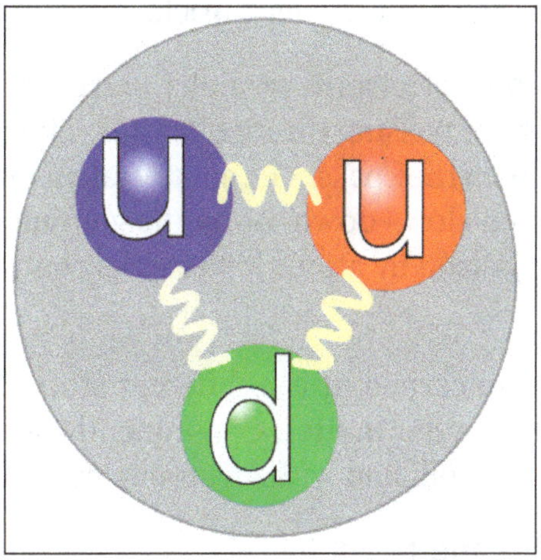

The quark structure of the neutron. There are two down quark and one up quark. The strong force is mediated by gluons. The strong force has three types of charges, the so called red, green and the blue. The color assignment of individual quarks is arbitrary, but all three colors must be present.

Properties

Outside the nucleus, neutrons are unstable and have a mean lifetime of 886 seconds (about 15 minutes), decaying by emitting an electron and antineutrino to become a proton. Neutrons in this unstable form are known as free neutrons. The same decay method (beta decay) occurs in some nuclei. Particles inside the nucleus are typically resonances between neutrons and protons, which transform into one another by the emission and absorption of pions. A neutron is classified as a baryon, and consists of two down quarks and one up quark. The neutron's antimatter equivalent is the antineutron.

Though most atoms have approximately equal numbers of protons and neutrons the number of neutrons can vary. Atoms of the same element, same number of protons, with different numbers of neutrons are called isotopes. The number of neutrons determines the isotope. For example, the carbon-12 isotope has six protons and six neutrons, while the carbon-14 isotope has six protons and eight neutrons.

Neutron Interactions

The neutron interacts through all four of the common classifications of physical interaction. These four are the electromagnetic, weak nuclear, strong nuclear, and gravitational interactions.

Although it is true that the neutron has zero net charge, it is nonetheless composed of electrically charged quarks, in the same way that a neutral atom is nonetheless composed of protons and electrons. As such, the neutron experiences the electromagnetic interaction. The net charge is zero, so if you are far enough away from the neutron that it appears to occupy no volume, then the total effect of the electric force will add up to zero. The movement of the charges inside the neutrons do not cancel however, and this is what gives the neutron its nonzero magnetic moment.

Gravity is often not discussed when talking about neutrons. This is because neutrons are usually studied in terms of subatomic interactions. In the subatomic world, gravity is undetectable relative to the other forces, which are much stronger. This having been said, a neutron accelerates at the same rate in the earth's gravitational field as a lead brick.

Charged particles (such as protons, electrons, or alpha particles) and electromagnetic radiation (such as gamma rays) lose energy in passing through matter. They exert electric forces that ionize atoms of the material through which they pass. The energy taken up in ionization equals the energy lost by the charged particle, which slows down, or by the gamma ray, which is absorbed or scattered. The neutron, by contrast, is seen by atoms it passes as containing no electric charge, and so does not create any ionization.

As far as the nuclear forces are concerned, it is a different story. Nuclear forces play the leading role when neutrons pass through regular matter. Consequently, a free neutron goes on its way unchecked until it makes a "head-on" collision with an atomic nucleus.

When this happens, the neutrons and target nuclei can be scattered (deflected or slowed down), absorbed, or transformed into something different. In the case of the reaction $n + {}^3He \rightarrow {}^1H + {}^3H$ (n: neutron; ${}^3He$: nucleus consisting of two protons and one neutron; ${}^1H$: nucleus consisting of a only proton; ${}^3H$: nucleus consisting of one proton and two neutrons). For example, the proton and the neutron appear to have exchanged places, and kinetic energy is released. In many cases, secondary particles are created and energy can be used up or released.

Neutrons, like other particles, can undergo elastic collisions. A collision is elastic under the special case where kinetic energy is conserved. Billiard balls for example typically undergo elastic collisions. The law of conservation of momentum also applies as it does for any collision. If the nucleus that is struck in an elastic collision is heavy, it acquires relatively little speed, but if it is a proton, which is approximately equal in mass to the neutron, it is projected forward with a large fraction of the original speed of the neutron, which is itself correspondingly slowed.

Neutron Detection

The common means of detecting a charged particle by looking for a track of ionization does not work for neutrons. The means that we have for detecting neutrons consists of allowing them to interact with atomic nuclei.

A common method for detecting neutrons involves converting the energy released from such reactions into electrical signals. The isotopes ^{3}He, ^{6}Li, ^{10}B, ^{233}U, ^{235}U and ^{239}Pu are useful for this purpose. These substances are mostly rare or tightly controlled, except for ^{10}B which is readily available.

The neutron plays an important role in many nuclear reactions. For example, neutron capture often results in neutron activation, inducing nuclear fission. In particular, knowledge of neutrons and their behavior has been important in the development of nuclear reactors and nuclear weapons.

The development of "neutron lenses" based on total internal reflection within hollow glass capillary tubes or by reflection from dimpled aluminum plates has driven ongoing research into neutron microscopy and neutron/gamma ray tomography.

One use of neutron emitters is the detection of light nuclei, particularly the hydrogen found in water molecules. When a fast neutron collides with a light nucleus, it loses a large fraction of its energy. By measuring the rate at which slow neutrons return to the probe after reflecting off of hydrogen nuclei, a neutron probe may determine the water content in soil.

Neutron Sources

Due to the fact that free neutrons are unstable, they (neutron radiation) can be obtained only from nuclear disintegrations or high-energy reactions (such as cosmic radiation or in accelerators). Free neutron beams are obtained from neutron sources by neutron transport.

Neutrons' lack of electric charge prevents engineers or experimentalists from being able to steer or accelerate them. Charged particles can be accelerated, decelerated, or deflected by electric or magnetic fields. However, these methods have almost no effect on neutrons (there is a small effect of a magnetic field on the free neutron because of its magnetic moment).

Antineutron

The antineutron is the antiparticle of the neutron. It was discovered by Bruce Cork in the year 1956, a year after the antiproton was discovered.

CPT-symmetry puts strong constraints on the relative properties of particles and antiparticles, and therefore is open to stringent tests. The masses of the neutron and antineutron are equal to one part in $(9 \pm 5) \times 10^{-5}$.

Atomic Physics

Atomic physics (or atom physics) is a field of physics that involves investigation of the structures of atoms, their energy states, and their interactions with other particles and electromagnetic radiation. In this field of physics, atoms are studied as isolated systems made up of nuclei and electrons.

Its primary concern is related to the arrangement of electrons around the nucleus and the processes by which these arrangements change. It includes the study of atoms in the form of ions as well as in the neutral state. For purposes of this discussion, it should be assumed that the term atom includes ions, unless otherwise stated. Through studies of the structure and behavior of atoms, scientists have been able to explain and predict the properties of chemical elements, and by extension, chemical compounds.

The term atomic physics is often associated with nuclear power and nuclear bombs, due to the synonymous use of atomic and nuclear in standard English. However, physicists distinguish between atomic physics, which deals with the atom as a system consisting of a nucleus and electrons, and nuclear physics, which considers atomic nuclei alone. As with many scientific fields, strict delineation can be highly contrived and atomic physics is often considered in the wider context of atomic, molecular, and optical physics.

Isolated Atoms

Atomic physics involves investigation of atoms as isolated entities. In atomic models, the atom is described as consisting of a single nucleus that is surrounded by one or more bound electrons. It is not concerned with the formation of molecules (although much of the physics is identical), nor does it examine atoms in a solid state as condensed matter. It is concerned with processes such as ionization and excitation by photons or collisions with atomic particles.

In practical terms, modeling atoms in isolation may not seem realistic. However, if one considers atoms in a gas or plasma, then the time scales for atom-atom interactions are huge compared to the atomic processes being examined here. This means that the individual atoms can be treated as if each were in isolation because for the vast majority of the time they are. By this consideration, atomic physics provides the underlying theory in plasma physics and atmospheric physics, although both deal with huge numbers of atoms.

Electronic Configuration

Electrons form notional shells around the nucleus. These electrons are naturally in their lowest energy state, called the ground state, but they can be excited to higher energy states by the absorption of energy from light (photons), magnetic fields, or interaction with a colliding particle (typically other electrons). The excited electron may still be bound to the nucleus, in which case they should, after a certain period of time, decay back to the original ground state. In so doing, energy is released as photons. There are strict selection rules regarding the electronic configurations that can be reached by excitation by light, but there are no such rules for excitation by collision processes.

If an electron is sufficiently excited, it may break free of the nucleus and no longer remain part of the atom. The remaining system is an ions and the atom is said to have been ionized, having been left in a charged state.

2

Models of Atomic Structure

A model which represents what the structure of an atom could look like on the basis of the knowledge of the behavior of atoms is known as model of atomic structure. There are a number of models which have been presented by different physicists such as Dalton, Thomson, Rutherford and Bohr. The topics elaborated in this chapter will help in gaining a better perspective about the models put forward by these physicists.

An atomic model represents what the structure of an atom could look like, based on what we know about how atoms behave. It is not necessarily a true picture of the exact structure of an atom. Models are often simplified. The small toy cars that you may have played with as a child are models. They give you a good idea of what a real car looks like, but they are much smaller and much simpler. A model cannot always be absolutely accurate and it is important that we realize this, so that we do not build up an incorrect idea about something.

Dalton's Atomic Model

Dalton's atomic model sets up the building blocks for others to improve on. Though some of his conclusions were incorrect, his contributions were vital. He defined an atom as the smallest indivisible particle.

Though we know today that they can be further divided into protons, neutrons, and electrons, his explanation was revolutionary for that period of time.

Basic Laws of Atomic Theory

Law of Conservation of Mass

The law of conservation of mass states that the net change in mass of the reactants and products before and after a chemical reaction is zero. This means mass can neither be created nor destroyed. In other words, the total mass in a chemical reaction remains constant.

This law was formulated by Antoine Lavoisier in 1789. It was later found to be slightly inaccurate, as in the course of chemical reactions mass can interconvert with heat and bond energy. However, these losses are very small, several orders of magnitude smaller than the mass of the reactants, so that this law is an excellent approximation.

Does the following chemical reaction obey the law of conservation of mass?

$$Ca(OH)_2 + CO_2 \rightarrow CaCO_3 + H_2O$$

The mass of Ca, O, H, and C are 40 u, 16 u, 1 u, and 12 u, respectively.

Yes, they obey the law of conservation of mass. Let's verify it. The molecular mass of,

$$\begin{aligned} Ca(OH)_2 &= 40+32+2 \\ &= 74 \\ CO_2 &= 12+32 \\ &= 44 \\ CaCO_3 &= 40+12+48 \\ &= 100 \\ H_2O &= 2+16 \\ &= 18. \end{aligned}$$

Substituting these values in the equation,

$$\begin{aligned} 74+44 &= 100+18 \\ 118 &= 118. \end{aligned}$$

Law of Constant Proportions

The law of constant proportions states that when a compound is broken, the masses of the constituent elements remain in the same proportion. Or in a chemical compound, the elements are always present in definite proportions by mass. It means each compound has the same elements in the same proportions, irrespective of where the compound was obtained, who prepared the compound, or the mass of the compound. This law was formulated and proven by Joseph Louis Proust in 1799.

For example, a person living in Australia sent a 100 ml sample of $CaCO_3$ (calcium carbonate) to a person living in India. The person living in India made his own sample of 200 ml and compared it to his friend's. Which of the two compounds has a greater ratio of Ca:C? Both contain equal ratio of Ca and C. This is guaranteed by the law of constant proportions.

Law of Multiple Proportions

The law of multiple proportions states that when two elements form two or more compounds between them, the ratio of the masses of the second element in each compound can be expressed in the form of small whole numbers. This law was proposed by John Dalton, and it is a combination of the previous laws.

Carbon combines with oxygen to form two different compounds (under different circumstances); one is the most common gas CO_2 and the other is CO. Do they obey the law of multiple proportions?

We know that the mass of carbon is 12 u and that of oxygen is 16 u. So, we can say that 12 g of carbon combines with 32 g of oxygen to form CO_2. Similarly, 12 g of carbon combines with 16 g of oxygen to form CO.

So, the ratio of oxygen in the first and second compound is $\frac{32}{16} = \frac{2}{1} = 2,$ which is a whole number.

There is one other law which was proposed to find the relation between two different compounds.

Law of Reciprocal Proportions

The law of reciprocal proportions states that when two different elements combine with the same quantity of a third element, the ratio in which they do so will be the same or a multiple of the proportion in which they combine with each other. This law was proposed by Jeremias Ritcher in 1792.

Dalton's Atomic Theory

Dalton picked up the idea of divisibility of matter to explain the nature of atoms. He studied the laws of chemical combinations carefully and came to a conclusion about the characteristics of atoms.

His statements were based on the three laws. He stated the following postulates about his atomic theory.

- Matter is made of very tiny particles called atoms.
- Atoms are indivisible structures, which can neither be created nor destroyed during a chemical reaction (based on the law of conservation of mass).
- All atoms of a particular element are similar in all respects, be it their physical or chemical properties.
- Inversely, atoms of different elements show different properties, and they have different masses and different chemical properties.
- Atoms combine in the ratio of small whole numbers to form stable compounds, which is how they exist in nature.
- The relative number and the kinds of atoms in a given compound are always in a fixed ratio (based on the law of constant proportions).

Drawbacks

- The first part of the second postulate was not accepted. Bohr's model proposed that the atoms could be further divided into protons, neutrons, and electrons.
- The third postulate was also proven to be wrong because of the existence of isotopes, which are atoms of the same element but of different masses.
- The fourth postulate was also proven to be wrong because of the existence of isobars, which are atoms of different elements but of the same mass.

Dalton's Model of an Atom

Based on all his observations, Dalton proposed his model of an atom. It is often referred to as the billiard ball model. He defined an atom to be a ball-like structure, as the concepts of atomic nucleus and electrons were unknown at the time. If you asked Dalton to draw the diagram of an atom, he would've simply drawn a circle.

Later, he tried to symbolize atoms, and he became one of the first scientists to assign such symbols. He gave a specific symbol to each atom.

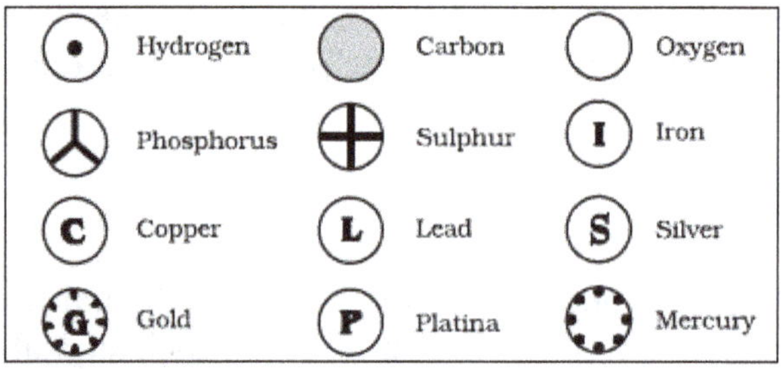

It was only after J.J. Thompson proposed his model that the true concepts had come into existence. Later, Rutherford worked on Dalton's and Thompson's models and brought out a roughly correct shape of the concept. Finally, Bohr's model and the quantum mechanical model gave a complete model which we know of today.

Thomson's Plum Pudding Model

Thomson atomic model was proposed by William Thomson in the year 1900. This model explained the description of an inner structure of the atom theoretically. It was strongly supported by Sir Joseph Thomson, who had discovered the electron earlier.

During cathode ray tube experiment, a negatively charged particle was discovered by J.J. Thomson. This experiment took place in the year 1897. Cathode ray tube is a vacuum tube. The negative particle was called an electron.

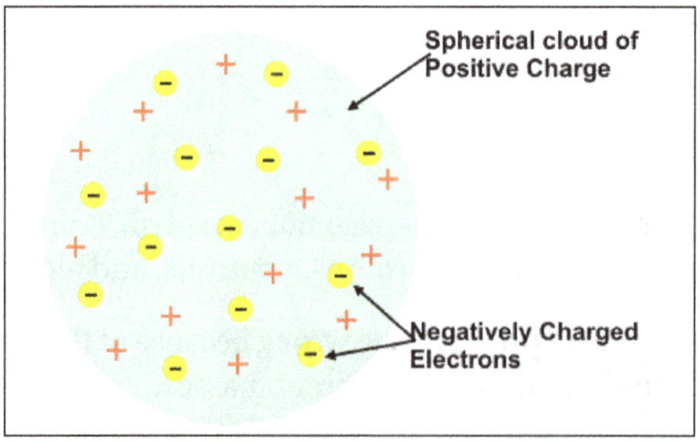

Thomson assumed that an electron is two thousand times lighter than a proton and believed that an atom is made up of thousands of electrons. In this atomic structure model, he considered atoms surrounded by a cloud having positive as well as negative charges. The demonstration of the ionization of air by X-ray was also done by him together with Rutherford. They were the first to demonstrate it. Thomson's model of an atom is similar to a plum pudding.

Postulates of Thomson's Atomic Model

Postulate 1: An atom consists of a positively charged sphere with electrons embedded in it.

Postulate 2: An atom as a whole is electrically neutral because the negative and positive charges are equal in magnitude.

Thomson atomic model is compared to watermelon. Where he considered:

- Watermelon seeds as negatively charged particles.
- The red part of the watermelon as positively charged.

Limitations of Thomson's Atomic Model

- It failed to explain the stability of an atom because his model of atom failed to explain how a positive charge holds the negatively charged electrons in an atom. Therefore, this theory also failed to account for the position of the nucleus in an atom.
- Thomson's model failed to explain the scattering of alpha particles by thin metal foils.
- No experimental evidence in its support.

Although Thomson's model was not an accurate model to account for the atomic structure, it proved to be the base for the development of other atomic models.

Rutherford Model

Rutherford proposed that an atom is composed of empty space mostly with electrons orbiting in a set, predictable paths around fixed, positively charged nucleus.

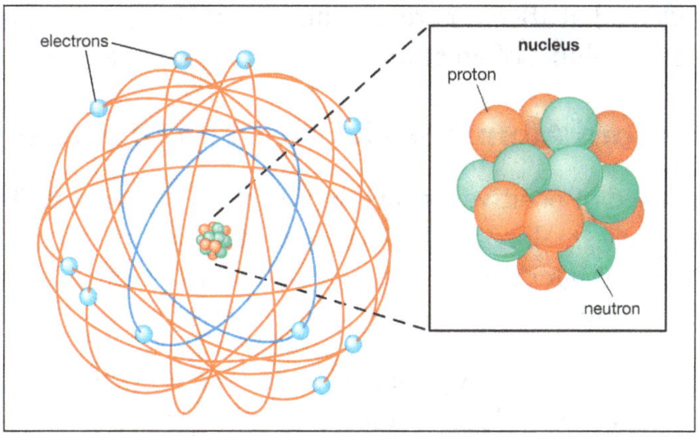

Rutherford's Atomic Model.

Rutherford Atomic Model Experiment

In Rutherford's experiment, he bombarded high energy streams of α-particles on a thin gold foil of 100 nm thickness. The streams of α-particles were directed from a radioactive source. He conducted the experiment to study the deflection produced in the trajectory of α-particles after interaction with the thin sheet of gold. To study the deflection, he placed a screen made up of zinc sulfide around the gold foil. The observations made by Rutherford contradicted the plum pudding model given by J.J. Thomson.

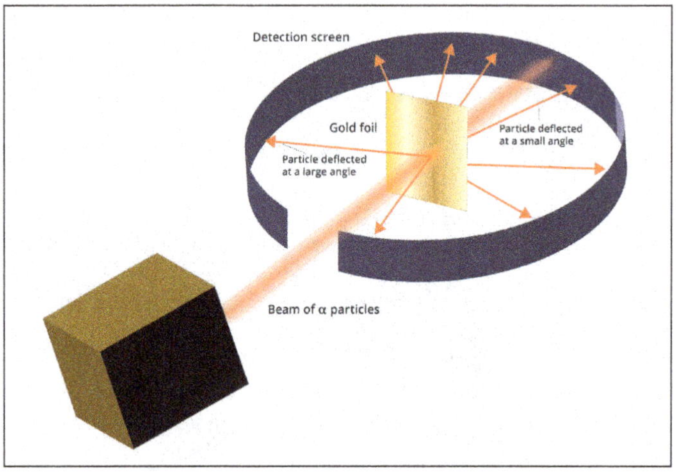

Rutherford's Gold Foil Experiment.

Observations of Rutherford Model Experiment

On the basis of the observations made during the experiment, Rutherford concluded that:

- Major space in an atom is empty – A large fraction of α-particles passed through the gold sheet without getting deflected. Therefore, the major part of an atom must be empty.
- The positive charge in an atom is not distributed uniformly and it is concentrated in a very small volume – Few α-particles when bombarded were deflected by the gold sheet. They were deflected minutely and at very small angles. Therefore he made the above conclusion.
- Very few α-particles had deflected at large angles or deflected back. Moreover, very few particles had deflected at 180°. Therefore, he concluded that the positively charged particles covered a small volume of an atom in comparison to the total volume of an atom.

Postulates of Rutherford Atomic Model Based on Observations

- An atom is composed of positively charged particles. Majority of the mass of an atom was concentrated in a very small region. This region of the atom was called as the nucleus of an atom. It was found out later that the very small and dense nucleus of an atom is composed of neutrons and protons.
- Atoms nucleus is surrounded by negatively charged particles called electrons. The electrons revolve around the nucleus in a fixed circular path at very high speed. These fixed circular paths were termed as "orbits."

- An atom has no net charge or they are electrically neutral because electrons are negatively charged and the densely concentrated nucleus is positively charged. A strong electrostatic force of attractions holds together the nucleus and electrons.
- The size of the nucleus of an atom is very small in comparison to the total size of an atom.

Limitations of Rutherford Atomic Model

Rutherford's experiment was unable to explain certain things. They are:

- Rutherford's model was unable to explain the stability of an atom. According to Rutherford's postulate, electrons revolve at a very high speed around a nucleus of an atom in a fixed orbit. However, Maxwell explained accelerated charged particles release electromagnetic radiations. Therefore, electrons revolving around the nucleus will release electromagnetic radiation.
- The electromagnetic radiation will have energy from the electronic motion as a result of which the orbits will gradually shrink. Finally, the orbits will shrink and collapse in the nucleus of an atom. According to the calculations, if Maxwell's explanation is followed Rutherford's model will collapse with 10^{-8} seconds. Therefore, Rutherford atomic model was not following Maxwell's theory and it was unable to explain an atom's stability.
- Rutherford's theory was incomplete because it did not mention anything about the arrangement of electrons in the orbit. This was one of the major drawbacks of Rutherford atomic model.

Even though the early atomic models were inaccurate and could not explain the structure of atom and experimental results properly. But it formed the basis of the quantum mechanics and helped the future development of quantum mechanics.

Bohr Model of Atom

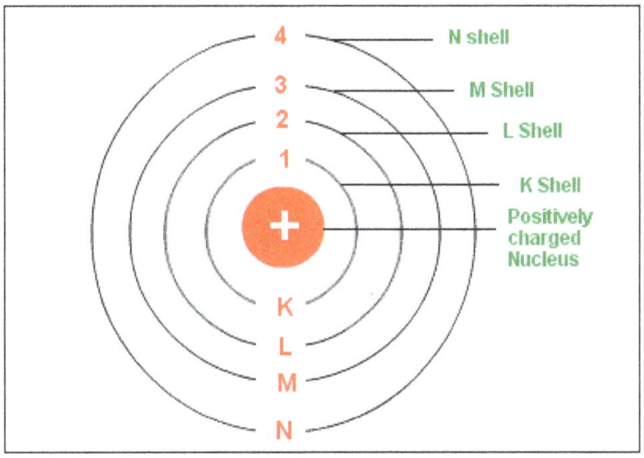

Bohr's Model of an Atom.

Bohr model of the atom was proposed by Neil Bohr in 1915. It came into existence with the modification of Rutherford's model of an atom. Rutherford's model introduced the nuclear model of an atom, in which he explained that a nucleus (positively charged) is surrounded by negatively charged electrons. Bohr modified this atomic structure model by explaining that electrons move in fixed orbital's (shells) and not anywhere in between and he also explained that each orbit (shell) has a fixed energy level. Rutherford basically explained nucleus of an atom and Bohr modified that model into electrons and their energy levels.

Bohr's model consists of a small nucleus (positively charged) surrounded by negative electrons moving around the nucleus in orbits. Bohr found that an electron located away from the nucleus has more energy, and electrons close to the nucleus have less energy.

Postulates of Bohr's Model of an Atom

- In an atom, electrons (negatively charged) revolve around the positively charged nucleus in a definite circular path called as orbits or shells.
- Each orbit or shell has a fixed energy and these circular orbits are known as orbital shells.
- The energy levels are represented by an integer (n = 1, 2, 3...) known as the quantum number. This range of quantum number starts from nucleus side with n = 1 having the lowest energy level. The orbits n = 1, 2, 3, 4... are assigned as K, L, M, N.... shells and when an electron attains the lowest energy level it is said to be in the ground state.
- The electrons in an atom move from a lower energy level to a higher energy level by gaining the required energy and an electron moves from a higher energy level to lower energy level by losing energy.

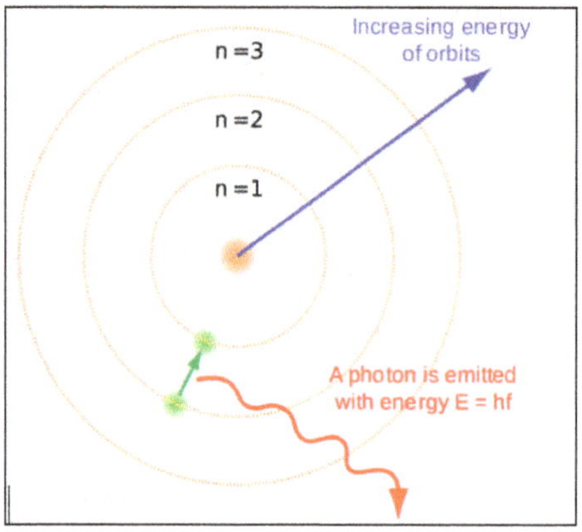

Limitations of Bohr's Model of an Atom

- Bohr's model of an atom failed to explain the Zeeman Effect (effect of magnetic field on the spectra of atoms).
- It also failed to explain the Stark effect (effect of electric field on the spectra of atoms).
- It violates the Heisenberg Uncertainty Principle.
- It could not explain the spectra obtained from larger atoms.

Sommerfeld Model of the Atom

In order to explain the observed fine structure of spectral lines, Sommerfeld introduced two main modifications in Bohr's theory:

- According to Sommerfeld, the path of an electron around the nucleus, in general, is an ellipse with the nucleus at one of its foci.
- The velocity of the electron moving in an elliptical orbit varies at different parts of the orbit. This causes the relativistic variation in the mass of the moving electron.

Now, when elliptical orbits are permitted, one has to deal with two variable quantities.

- The varying distance of the electron from the nucleus (r).
- The varying angular position of the electron with respect to the nucleus i.e the azimuthal angle φ.

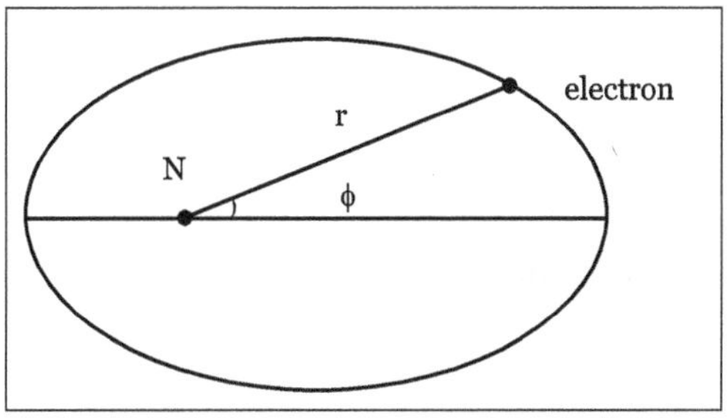

Sommerfeld atom model.

To deal with these two variables, two quantum numbers are introduced:

- The principal quantum number n of Bohr's theory, which determines the energy of the electrons,
- A new quantum number called orbital (or azimuthal) quantum number (l) which has been introduced to characterize the angular momentum in an orbit i.e., it determines the orbital angular momentum of the electron. Its values vary from zero to (n-1) in steps of unity.

This orbital quantum number (l) is useful in finding the possible elliptical orbits. The possible elliptical orbits are such that:

$$b/a = l+1/n$$

where a and b are semi-major and semi-minor axes respectively of the ellipse.

According to Sommerfeld's model, for any principal quantum number n, there are n possible orbits of varying eccentricities called sub-orbits or sub-shells. Out of n subshells, one is circular and the remaining (i.e., n-1) are elliptical in shape.

These possible sub-orbits possess slightly different energies because of the relativistic variation of the electron mass.

Consider the first energy level (n = 1). When n = 1, l = 0 i.e., in this energy level, there is only one orbit or sub-shell for the electron. Also, when $a = b$, the two axes of the ellipse are equal. As a result of this, the orbit corresponding to n = 1 is circular. This subshell is designated as s sub-shell. Since, this sub-shell belongs to n = 1, it is designated as 1s.

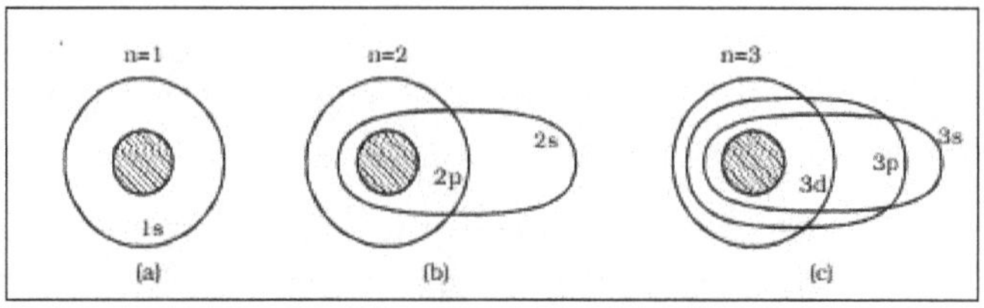

Various sub-shells for the electrons.

Similarly, for the second energy level n = 2, there are two permissible sub-shells for the electrons. For n = 2, l can take two values, 0 and 1.

When, $n = 2, l = 0$.

$$b/a = 0+1/2 = 1/2$$

or

$$b = a/2$$

This subshell corresponding to l = 0 is elliptical in shape and is designated as 2s,

when $n = 2, l = 1$.

$$b/a = 1+1/2 = 2/2 = 1$$

or

$$b = a$$

This sub-shell corresponding to l = 1 is circular in shape and is designated as 2p.

For n = 3, l has three values 0, 1 and 2, i.e. there are three permissible sub-shells for the electrons.

When, $n = 3, l = 0$.

$$b/a = (0+1)/3 = 1/3 = 1 \text{ or } b = a/3$$

When, $n = 3, l = 2$.

$$b/a = (1+1)/3 = 2/3 = 1 \text{ or } b = 2a/3$$

and when, $n = 3, l = 2$.

$$b/a = (2+1)/3 = 3/3 = 1 \text{ or } b = a$$

The sub-shells corresponding to l = 0, 1 and 2 are designated as 3s, 3p and 3d respectively. The circular shell is designated as 3d and the other two are elliptical in shape.

It is common practice to assign letters to *l*-values as given below:

- Orbital quantum number l : 0 1 2 3 4
- Electron state : *s* *p* *d* *f* *g*

Hence, electrons in the l = 0, 1, 2, 3, 4 states are said to be in the *s, p, d, f, g* states.

Fine Structure of Spectral Line

Based on Sommerfeld atom model, the total energy of an electron in the elliptical orbit can be shown as,

$$En = (-me^4Z^2) \; / \; (8\varepsilon_0^2 h^2 n^2)$$

This expression is the same as that obtained by Bohr. Thus the introduction of elliptical orbits gives no new energy levels and hence no new transition. In this way, the attempt of Sommerfeld to explain the fine structure of spectral lines failed. But soon, on the basis of variation of mass of electron with velocity, Sommerfeld could find the solution for the problem of the fine structure of the spectral lines.

According to Sommerfeld, the velocity of the electron is maximum when the electron is nearest to the nucleus and minimum when it is farthest from the nucleus, since the orbit of the electron is elliptical. This implies that the effective mass of the electron will be different at different parts of its orbit. Taking into account the relativistic variation of the mass of the electron, Sommerfeld modified his theory and showed that the path of electron is not a simple ellipse but a precessing ellipse called a rosette.

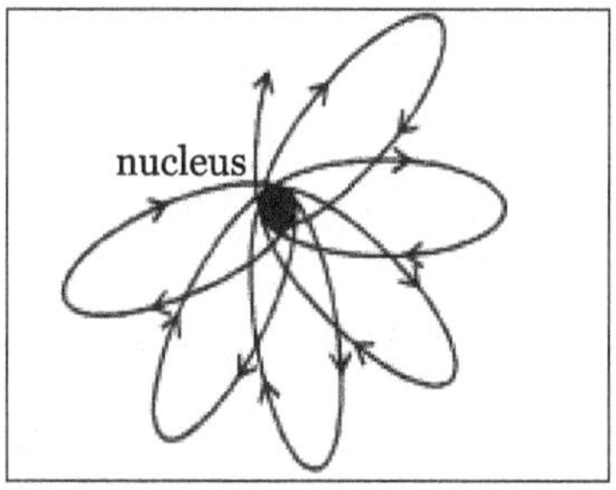

Rosette path of an electron.

Based on this idea, Sommerfeld successfully explained the fine structure of spectral lines of hydrogen atom.

Drawbacks Sommerfeld Model of the Atom

- Though Sommerfeld's modification gave a theoretical background of the fine structure of spectral lines of hydrogen, it could not predict the correct number of observed fine structure of these lines.
- It could not explain the distribution and arrangement of electrons in atoms.
- Sommerfeld's model was unable to explain the spectra of alkali metals such as sodium, potassium etc.

- It could not explain Zeeman and Stark effect.
- This model does not give any explanation for the intensities of the spectral lines.

Quantum Mechanical Model of Atom

A major problem with Bohr's model was that it treated electrons as particles that existed in precisely-defined orbits. Based on de Broglie's idea that particles could exhibit wavelike behavior, Austrian physicist Erwin Schrödinger theorized that the behavior of electrons within atoms could be explained by treating them mathematically as matter waves. This model, which is the basis of the modern understanding of the atom, is known as the quantum mechanical or wave mechanical model.

The fact that there are only certain allowable states or energies that an electron in an atom can have is similar to a standing wave. You are probably already familiar with standing waves from stringed musical instruments. For example, when a string is plucked on a guitar, the string vibrates in the shape of a standing wave.

Notice that there are points of zero displacement or nodes that occur along the standing wave. The nodes are marked with red dots. Since the string in the animation is fixed at both ends, this leads to the limitation that only certain wavelengths are allowed for any standing wave. As such, the vibrations are quantized.

Schrödinger's Equation

On a very simple level, we can think of electrons as standing matter waves that have certain allowed energies. Schrödinger formulated a model of the atom that assumed the electrons could be treated at matter waves. The basic form of Schrödinger's wave equation is as follows:

$$\hat{H}\Psi = E\Psi$$

Ψ is called a wave function; $\hat{H}$ is known as the Hamiltonian operator; and E is the binding energy of the electron. Solving Schrödinger's equation yields multiple wave functions as solutions, each with an allowed value for E.

Interpreting exactly what the wave functions tell us is a bit tricky. Due to the Heisenberg uncertainty principle, it is impossible to know for a given electron both its position and its energy. Since knowing the energy of an electron is necessary for predicting the chemical reactivity of an atom, chemists generally accept that we can only approximate the location of the electron.

How do chemists approximate the location of the electron? The wave functions that are derived from Schrödinger's equation for a specific atom are also called atomic orbitals. Chemists define an atomic orbital as the region within an atom that encloses where the electron is likely to be 90% of the time.

Orbitals and Probability Density

The value of the wave function Ψ at a given point in space x, y, z, is proportional to the amplitude of the electron matter wave at that point. However, many wave functions are complex functions containing $i = \sqrt{-1}$, and the amplitude of the matter wave has no real physical significance.

Luckily, the square of the wave function, Ψ^2 is a little more useful. This is because the square of a wave function is proportional to the probability of finding an electron in a particular volume of space within an atom. The function Ψ^2 is often called the probability density.

The probability density for an electron can be visualized in a number of different ways. For example, Ψ^2 can be represented by a graph in which varying intensity of color is used to show the relative probabilities of finding an electron in a given region in space. The greater the probability of finding an electron in a particular volume the higher the density of the color in that region. The image below shows the probability distributions for the spherical 1s, 2s, and 3s orbitals.

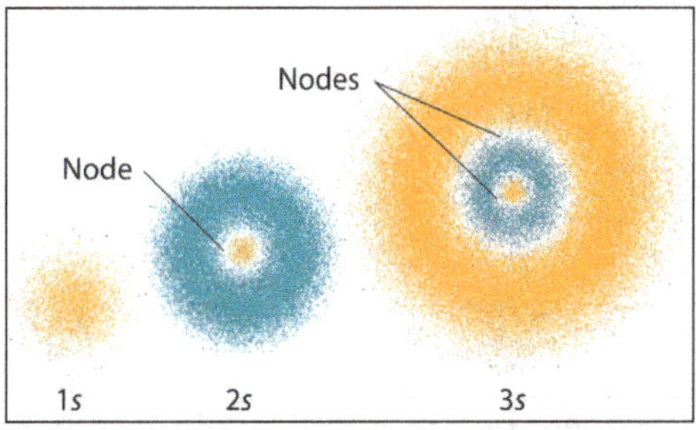

Probability distributions for 1s, 2s, and 3s orbitals. Greater color intensity indicates regions where electrons are more likely to exist. Nodes indicate regions where an electron has zero probability of being found.

Notice that the 2s and 3s orbitals contain nodes—regions in which an electron has a 0% probability of being found. The existence of nodes is analogous to the standing waves. The alternating colors in the 2s and 3s orbitals represent regions of the orbital with different phases, which is an important consideration in chemical bonding.

Another way of picturing probabilities for electrons in orbitals is by plotting the surface density as a function of the distance from the nucleus, r.

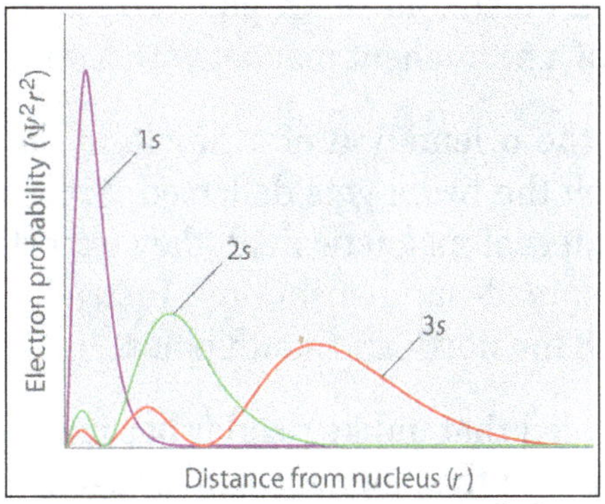

A radial probability graph showing surface probability $\Psi^2 r^2$ vs. r. Electrons occupying higher-energy orbitals have greater probabilities of being found farther from the nucleus.

The surface density is the probability of finding the electron in a thin shell with radius r. This is called a radial probability graph. On the left is a radial probability graph for the 1s, 2s, and 3s orbitals. Notice that as the energy level of the orbital increases from 1s to 2s to 3s, the probability of finding an electron farther from the nucleus increases as well.

Shapes of Atomic Orbitals

So far we have been examining s orbitals, which are spherical. As such, the distance from the nucleus, *r*, is the main factor affecting an electron's probability distribution. However, for other types of orbitals such as p, d, and f orbitals, the electron's angular position relative to the nucleus also becomes a factor in the probability density. This leads to more interesting orbital shapes, such as the ones in the following image.

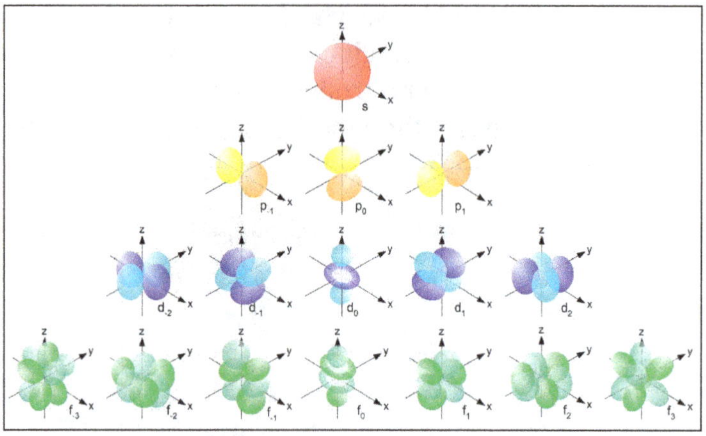

Schematics showing the general shapes of s, p, d, and f orbitals.

The p orbitals are shaped like dumbbells that are oriented along one of the axes–x, y, z. The d orbitals can be described as having a clover shape with four possible orientations—with the exception of the d orbital that almost looks like a p orbital with a donut going around the middle. It's not even worth attempting to describe the f orbitals.

Electron Spin: The Stern-Gerlach Experiment

In 1922, German physicists Otto Stern and Walther Gerlach hypothesized that electrons behaved as tiny bar magnets, each with a north and south pole. To test this theory, they fired a beam of silver atoms between the poles of a permanent magnet with a stronger north pole than south pole.

According to classical physics, the orientation of a dipole in an external magnetic field should determine the direction in which the beam gets deflected. Since a bar magnet can have a range of orientations relative to the external magnetic field, they expected to see atoms being deflected by different amounts to give a spread-out distribution. Instead, Stern and Gerlach observed the atoms were split cleanly between the north and south poles.

These experimental results revealed that unlike regular bar magnets, electrons could only exhibit two possible orientations: either with the magnetic field or against it. This phenomenon, in which electrons can exist in only one of two possible magnetic states, could not be explained using classical physics. Scientists refer to this property of electrons as electron spin: any given electron is either spin-up or spin-down. We sometimes represent electron spin by drawing electrons as arrows pointing up, ↑ or down, ↓.

One consequence of electron spin is that a maximum of two electrons can occupy any given orbital, and the two electrons occupying the same orbital must have opposite spin. This is also called the Pauli exclusion principle.

Electron Configuration

Electron configuration is the distribution of electrons of an atom or molecule (or other physical structure) in atomic or molecular orbitals; for example, the electron configuration of a neon atom is $1s^2\ 2s^2\ 2p^6$. Electronic configurations describe electrons as each moving independently in an orbital, in an average field created by all other orbitals. From electron configuration, an atoms' reactivity and potential for corrosion can be determined.

Knowledge of the electron configuration of different atoms is useful in understanding the structure of the periodic table of elements. The concept is also useful for describing the chemical bonds that hold atoms together. In bulk materials, this idea helps explain the peculiar properties of lasers and semiconductors.

The electron configuration of an atom describes the orbitals occupied by electrons on the atom. The basis of this prediction is a rule known as the Aufbau principle, which assumes that electrons are added to an atom, one at a time, starting with the lowest energy orbital, until all of the electrons have been placed in an appropriate orbital.

The electron configuration is used to describe the orbitals of an atom in its ground state, but it can also be used to represent an atom that has ionized into a cation or anion by compensating with the loss of or gain of electrons in their subsequent orbitals. Many of the physical and chemical properties of elements can be correlated to their unique electron configurations.

The most widespread application of electron configurations is in the rationalization of chemical properties, in both inorganic and organic chemistry. In effect, electron configurations, along with some simplified form of molecular orbital theory, have become the modern equivalent of the valence concept, describing the number and type of chemical bonds that an atom can be expected to form. A fundamental application of electron configurations is in the interpretation of atomic spectra.

Quantum Numbers

Quantum numbers are numbers assigned to all the electrons in an atom and they describe certain characteristics of the electron. It is very important to understand Quantum Numbers in order to understand the Structure of Atom.

An atom consists of a large number of orbitals which are distinguished from each other on the basis of their shape, size and orientation in space. The orbital characteristics are used to define the state of an electron completely and are expressed in terms of three numbers as stated, Principal quantum number, Azimuthal quantum number and Magnetic quantum number and Spin Quantum number.

Quantum numbers are those numbers that designate and distinguish various atomic orbitals and electrons present in an atom. A set of four numbers through which we can get the complete information about all the electrons in an atom, be it energy, location, space, type of orbital occupied, and even the orientation of that orbital is called Quantum Numbers.

Principal Quantum Number

The Principal Quantum Number represents the principal energy level or shell in which an electron revolves around the nucleus. It is denoted by the letter n and can have any integral value except the 0, i.e. n = 1, 2, 3, 4 … ….etc. The energies of the various principal shells will follow the sequence as:

K < L < M < N < O.....

1 < 2 < 3 < 4 < 5.......

Azimuthal Quantum Number

Azimuthal quantum number, also known as orbital quantum number determines the subshell to which an electron belongs. As a matter of result, the number of electronic jump increases and the number of lines at the same time.

- For a given value of n, it can have any integral value ranging from 0 to n – 1.
- For the 1st Shell, say K, n =1, you can have only one value i.e. l = 0.
- For the 2nd Shell, say L, n = 2, you can have two values i.e. l = 0 and 1.
- For the 3rd Shell, say M, n = 3, you can have three values i.e. l = 0, 1 and 2.
- For the 4th shells, say N, n = 4, you can have 4 values i.e. l = 0, 1, 2 and 3.

Magnetic Quantum Number

Magnetic Quantum Number denoted by the symbol m is what represents the orientation of atomic orbital in space. The value of the Magnetic Quantum Number, m, depends on the value of l. Magnetic Quantum Number can have a total number of (2l + 1).

Sublevel	l	m_l
s	0	0
p	1	-1, 0, +1
d	2	-2, -1, 0, +1, +2
f	3	-3, -2, -1, 0, +1, +2, +3

Spin Quantum Number

Spin Quantum Number represents the direction of the spin of the electrons. This can either be in the direction of clockwise or even anti-clockwise. Spin Quantum Number is denoted by the symbol s. It can have about only two values i.e. +1/2 or -1/2.

Aufbau Principle

The Aufbau principle dictates the manner in which electrons are filled in the atomic orbitals of an atom in its ground state. It states that electrons are filled into atomic orbitals in the increasing order of orbital energy level. According to the Aufbau principle, the available atomic orbitals with the lowest energy levels are occupied before those with higher energy levels.

The word 'Aufbau' has German roots and can be roughly translated as 'construct' or 'build up'. A diagram illustrating the order in which atomic orbitals are filled is provided below. Here, 'n' refers to the principal quantum number and 'l' is the azimuthal quantum number.

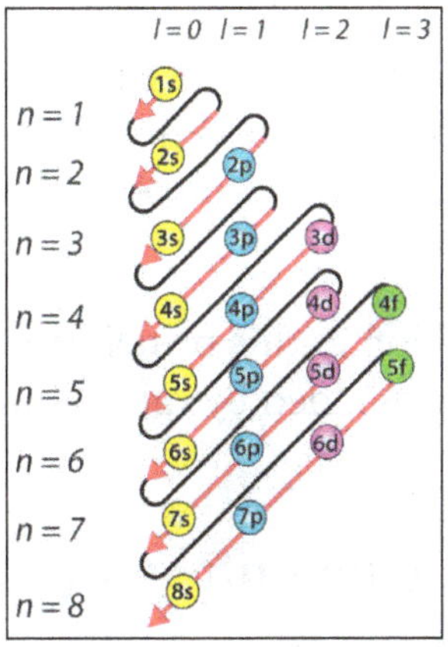

The Aufbau principle can be used to understand the location of electrons in an atom and their corresponding energy levels. For example, carbon has 6 electrons and its electronic configuration is $1s^2\ 2s^2\ 2p^2$.

It is important to note that each orbital can hold a maximum of two electrons (as per the Pauli exclusion principle). Also, the manner in which electrons are filled into orbitals in a single subshell must follow Hund's rule, i.e. every orbital in a given subshell must be singly occupied by electrons before any two electrons pair up in an orbital.

According to the Aufbau principle, electrons first occupy those orbitals whose energy is the lowest. This implies that the electrons enter the orbitals having higher energies only when orbitals with lower energies have been completely filled. The order in which the energy of orbitals increases can be determined with the help of the (n+l) rule, where the sum of the principal and azimuthal quantum numbers determines the energy level of the orbital. Lower (n+l) values correspond to lower orbital energies. If two orbitals share equal (n+l) values, the orbital with the lower n value is said to have lower energy associated with it.

The order in which the orbitals are filled with electrons is: 1s, 2s, 2p, 3s, 3p, 4s, 3d, 4p, 5s, 4d, 5p, 6s, 4f, 5d, 6p, 7s, 5f, 6d, 7p, and so on.

Exceptions

The electron configuration of chromium is $[Ar]3d^5 4s^1$ and not $[Ar]3d^4 4s^2$ (as suggested by the Aufbau principle). This exception is attributed to several factors such as the increased stability provided by half-filled subshells and the relatively low energy gap between the 3d and the 4s subshells.

The energy gap between the different subshells is illustrated below.

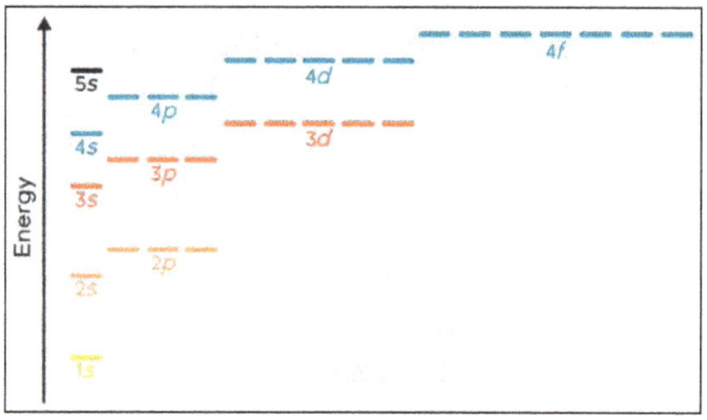

Half filled subshells feature lower electron-electron repulsions in the orbitals, thereby increasing the stability. Similarly, completely filled subshells also increase the stability of the atom. Therefore, the electron configurations of some atoms disobey the Aufbau principle (depending on the energy gap between the orbitals).

For example, copper is another exception to this principle with an electronic configuration corresponding to $[Ar]3d^{10}4s^1$. This can be explained by the stability provided by a completely filled 3d subshell.

Electronic Configuration using the Aufbau Principle

Writing the Electron Configuration of Sulfur:

- The atomic number of sulfur is 16, implying that it holds a total of 16 electrons.
- As per the Aufbau principle, two of these electrons are present in the 1s subshell, eight of them are present in the 2s and 2p subshell, and the remaining are distributed into the 3s and 3p subshells.
- Therefore, the electron configuration of sulfur can be written as $1s^2\ 2s^2\ 2p^2\ 3s^2\ 3p^6$.

Writing the Electron Configuration of Nitrogen:

- The element nitrogen has 7 electrons (since its atomic number is 7).
- The electrons are filled into the 1s, 2s, and 2p orbitals.
- The electron configuration of nitrogen can be written as $1s^2\ 2s^2\ 2p^3$.

Pauli Exclusion Principle

The Pauli exclusion principle says that every electron must be in its own unique state. In other words, no electrons in an atom are permitted to have an identical set of quantum numbers. The Pauli exclusion principle sits at the heart of chemistry, helping to explain the electron arrangements in atoms and molecules, and helping to rationalize patterns in the periodic table.

Every electron in an atom can be defined completely by four quantum numbers:

- n: The principal quantum number.

- l: The orbital angular momentum quantum number.
- m_l: The magnetic quantum number.
- m_s: The spin quantum number.

Example of the Pauli Exclusion Principle.

Consider argon's electron configuration:

$1s^2\ 2s^2\ 2p^6\ 3s^2\ 3p^6$

The exclusion principle asserts that every electron in an argon atom is in a unique state:

- The 1s level can accommodate two electrons with identical n, l, and m_l quantum numbers. Argon's pair of electrons in the 1s orbital satisfy the exclusion principle because they have opposite spins, meaning they have different spin quantum numbers, m_s. One spin is +½, the other is -½. (Instead of saying +½ or -½ often the electrons are said to be spin-up ↑ or spin-down ↓.)

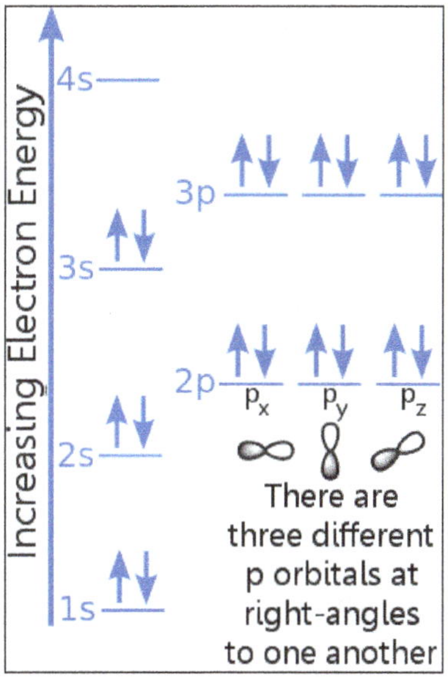

- The 2s level electrons have a different principal quantum number to those in the 1s orbital. The pair of 2s electrons differ from each other because they have opposite spins.
- The 2p level electrons have a different orbital angular momentum number from those in the s orbitals, hence the letter p rather than s. There are three p orbitals of equal energy, the p_x, p_y and p_z. These orbitals are different from one another because they have different orientations in space. Each of the p_x, p_y and p_z orbitals can accommodate a pair of electrons with opposite spins.
- The 3s level rises to a higher principal quantum number; this orbital accommodates an electron pair with opposite spins.
- The 3p level's description is similar to that for 2p, but the principal quantum number is higher: 3p lies at a higher energy than 2p.

Hund's Rules

Aufbau principle tells us that the lowest energy orbitals get filled by electrons first. After the lower energy orbitals are filled, the electrons move on to higher energy orbitals. The problem with this rule is that it does not tell about the three 2p orbitals and the order that they will be filled in. According to Hund's rule:

- Before the double occupation of any orbital, every orbital in the sub level is singly occupied.
- For the maximization of total spin, all electrons in a single occupancy orbital have the same spin.

An electron will not pair with another electron in a half-filled orbital as it has the ability to fill all its orbitals with similar energy. A large number of unpaired electrons are present in atoms which are at the ground state. If two electrons come in contact they would show the same behaviour as two magnets do. The electrons first try to get as far away from each other as possible before they have to pair up.

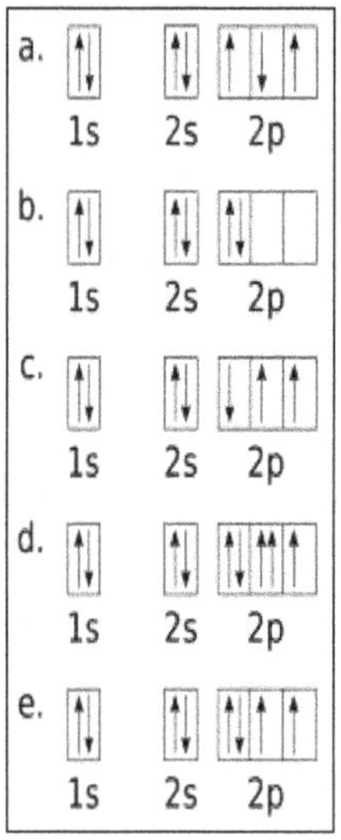

Hund's Rule.

Statement of Hund's Rule

It states that:

- In a sublevel, each orbital is singly occupied before it is doubly occupied.
- The electrons present in singly occupied orbitals possess identical spin.

Explanation of Hund's Rule

The electrons enter an empty orbital before pairing up. The electrons repel each other as they are negatively charged. The electrons do not share orbitals to reduce repulsion.

When we consider the second rule, the spins of unpaired electrons in singly occupied orbitals is the same. The initial electrons spin in the sub level decides what the spin of the other electrons would be. For instance, a carbon atom's electron configuration would be $1s^2\ 2s^2\ 2p^2$. The same orbital will be occupied by the two 2s electrons although different orbitals will be occupied by the two 2p electrons in reference to Hund's rule.

Hund's Rule of Maximum Multiplicity

The rule states that, for a stated electron configuration, the greatest value of spin multiplicity has the lowest energy term. It says if two or more than two orbitals having the same amount of energy are unoccupied then the electrons will start occupying them individually before they fill them in pairs. It is a rule which depends on the observation of atomic spectra, that is helpful in predicting the ground state of a molecule or an atom with one or more than one open electronic shells. This rule was discovered in the year 1925 by Friedrich Hund.

Writing Electron Configurations

You can start with the Periodic Table; that is the easiest way to determine electron configuration.

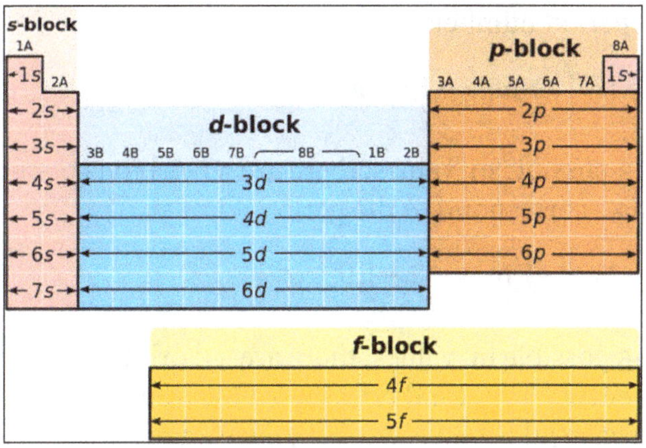

- Groups 1 and 2 are known as the s block, because those elements have outermost electrons in an s orbital.
- The p-block elements have their outermost electrons in the p orbital(s) but also have s orbital electrons.
- The d-block elements fill d orbitals with electrons before filling p orbitals (if at all).
- The f-block elements also fill f orbitals before filling p orbitals and before d orbitals.

Each orbital can hold a maximum of 2 electrons, which means s orbitals can go up to 2, p orbitals up to 6 (because 2 electrons max for 3 orbitals), d orbitals can go up to 10, and f orbitals can go up to 14. The number of electrons in each orbital is denoted in the superscript in the electron configuration.

The first quantum number is the principal quantum number n, which tells you the energy of those electrons. n ranges from 1 to 7 (theoretically can go 8+).

The second quantum number, the angular momentum l, ranges from 0 to $n-1$.

- l = 0 corresponds to an *s* orbital.
- l = 1 corresponds to a *p* orbital.
- l = 2 corresponds to a *d* orbital.
- l = 3 corresponds to an *f* orbital.
- l = 4 corresponds to an *g* orbital.

The third number, the magnetic quantum number m_l, tells you how many orbitals are in each sub-shell. They range from to l. For example, an electron in an s orbital on would have an l of 0 and therefore can have only m_l value of 0. There is only 1 possible m_l so there can only be 1 s orbital per energy level.

$$\text{s orbital} \rightarrow l = 0 \rightarrow m_l = 0$$

- Because there is only 1 possible m_l value, there is only 1 s orbital per energy level which also correspond to the period number.

$$\text{p orbital} \rightarrow l = 1 \rightarrow m_l = -1, 0, \text{ or } 1$$

- Because there are 3 possible m_l values, there are 3 p orbitals per energy level (starting from period 2), which also correspond to the period number.

$$\text{d orbital} \rightarrow l = 2 \rightarrow m_l = -2, -1, 0, 1, \text{or } 2$$

- Because there are 5 possible m_l values, there are 5 d orbitals per energy level (starting from period 4), but the energy level for d orbitals is 1 less than the period number, which is why period 4 elements have 3d orbitals, period 5 have 4d orbitals, etc.

$$\text{f orbital} \rightarrow l = 3 \rightarrow m_l = -3, -2, -1, 0, 1, 2, \text{ or } 3$$

- Because there are 7 possible m_l values, there are 7 f orbitals per energy level (starting from period 6), but the energy level for f orbitals is 2 less than the period number, hence the 4f orbitals in period 6 and 5f orbitals in period 7.

The final quantum number, magnetic spin m_s, tells you the direction in which an electron spins within an orbital, and can only have the values $+\frac{1}{2}$ for spin up and $-\frac{1}{2}$ for spin down.

Examples:

- H (Hydrogen) $1s^1$:

  Hydrogen only has 1 electron in the s orbital of the first energy level.

- He (Helium) $1s^2$:

  Helium has 2 electrons in the s orbital, which is now full, of the first energy level.

- N (Nitrogen) $1s^2\ 2s^2\ 3p^3$ or $[He]2s^2\ 2p^3$

  Nitrogen has 2 electrons in the s orbital of the first energy level, 2 more electrons in another s orbital on the 2nd energy level, and 3 electrons in the p orbitals of the 2nd energy level. The second representation is a common equivalent shorthand using the electron configuration of the noble gas that comes before the element.

- Br (Bromine) $1s^2\ 2s^2\ 2p^6\ 3s^2\ 3p^6\ 4s^2\ 3d^{10}\ 4p^5$ or $[Ar]4s^2\ 3d^{10}\ 4p^5$:

  Similar format as before, but notice how the 3d electrons are written before the 4p electrons. Also notice how much simpler and quicker it is to write the shorthand, especially in larger elements such as Bromine.

- U (Uranium)

  $1s^2 2s^2 2p^6 3s^2 3p^6 4s^2 3d^{10} 4p^6 5s^2 4d^{10} 5p^6 6s^2 4f^{14} 5d^{10} 6p^6 5f^3 6d^1 7s^2$ or $[Rn]5f^3 6d^1 7s^2$:

  As you can tell, the very large elements get very complex in their electron configurations, and the order of each subshell may not be correct (or may have many correct forms).

Notice in these examples that the superscripts for s orbitals only go up to 2, for p got up to 6, for d go up to 10, and for f go up to 14. This is due to the number of possible m_1 values for that orbital type.

Periodic trends are very useful when identifying and classifying elements.

References

- Daltons-atomic-model: brilliant.org, Retrieved 11 March, 2019
- Thomsons-model, chemistry: byjus.com, Retrieved 18 May, 2019
- Rutherfords-model-of-an-atom, structure-of-atom, chemistry: toppr.com, Retrieved 9 August, 2019
- Bohrs-model, chemistry: byjus.com, Retrieved 17 January, 2019
- Sommerfeld-atom-model-and-its-Drawbacks-2927: brainkart.com, Retrieved 14 June, 2019
- The-quantum-mechanical-model-of-the-atom, quantum-numbers-and-orbitals, quantum-physics, physics, science: khanacademy.org, Retrieved 20 January, 2019
- Electron-configuration: corrosionpedia.com, Retrieved 8 July, 2019
- Quantum-numbers, structure-of-atom, chemistry: toppr.com, Retrieved 20 February, 2019
- Aufbau-principle, chemistry: byjus.com, Retrieved 28 April, 2019
- Pauli-exclusion-principle: chemicool.com, Retrieved 13 February, 2019
- Hunds-rule, chemistry: byjus.com, Retrieved 10 April, 2019

3

The Hydrogen Atom

The atom of the chemical element hydrogen is called hydrogen atom. In its electrically neutral form, it contains a single proton and electron. Some of its isotopes are Deuterium and Tritium. This chapter has been carefully written to provide an easy understanding of the varied facets of hydrogen atom as well as its different models.

The hydrogen atom is the simplest atom in nature and therefore, a good starting point to study atoms and atomic structure. The hydrogen atom consists of a single negatively charged electron that moves about a positively charged proton.

Hydrogen (chemical symbol H, atomic number 1) is the lightest chemical element and the most abundant of all elements, constituting roughly 75 percent of the elemental mass of the universe. Stars in the main sequence are mainly composed of hydrogen in its plasma state.

In the Earth's natural environment, free (uncombined) hydrogen is relatively rare. At standard temperature and pressure, it takes the form of a colorless, odorless, tasteless, highly flammable gas made up of diatomic molecules (H_2). On the other hand, the element is widely distributed in combination with other elements, and many of its compounds are vital for living systems. Its most familiar compound is water (H_2O).

Elemental hydrogen is industrially produced from hydrocarbons such as methane, after which most elemental hydrogen is used "captively" (meaning locally, at the production site). The largest markets are about equally divided between fossil fuel upgrading (such as hydrocracking) and ammonia production (mostly for the fertilizer market).

The most common naturally occurring isotope of hydrogen, known as protium, has a single proton and no neutrons. In ionic compounds, it can take on either a positive charge (becoming a cation, H^+, which is a proton) or a negative charge (becoming an anion, H^-, called a hydride). It plays a particularly important role in acid-base chemistry, in which many reactions involve the exchange of protons between soluble molecules. As the only neutral atom for which the Schrödinger equation can be solved analytically, study of the energetics and bonding of the hydrogen atom has played a key role in the development of quantum mechanics.

Natural Occurrence

Hydrogen is the most abundant element in the universe, making up 75 percent of normal matter by mass and over 90 percent by number of atoms. This element is found in great abundance in stars and gas giant planets. Molecular clouds of H_2 are associated with star formation. Hydrogen plays a vital role in powering stars through proton-proton reaction nuclear fusion.

Throughout the universe, hydrogen is mostly found in the atomic and plasma states whose properties are quite different from molecular hydrogen. As a plasma hydrogen's electron and proton are not bound together, resulting in very high electrical conductivity and high emissivity (producing the light from the sun and other stars). The charged particles are highly influenced by magnetic and electric fields. For example, in the solar wind they interact with Earth's magnetosphere giving rise to Birkeland currents and the aurora. Hydrogen is found in the neutral atomic state in the Interstellar medium. The large amount of neutral hydrogen found in the damped Lyman-alpha systems is thought to dominate the cosmological baryonic density of the universe up to redshift $z = 4$.

A giant region of ionized hydrogen in the Triangulum Galaxy.

Under ordinary conditions on Earth, elemental hydrogen exists as the diatomic gas, H_2. However, hydrogen gas is very rare in the Earth's atmosphere (1 part per million by volume) because of its light weight, which enables it to escape Earth's gravity more easily than heavier gases. Although H atoms and H_2 molecules are abundant in interstellar space, they are difficult to generate, concentrate and purify on Earth. Still, hydrogen is the third most abundant element on the Earth's surface. Most of the Earth's hydrogen is in the form of chemical compounds such as hydrocarbons and water. Hydrogen gas is produced by some bacteria and algae and is a natural component of flatus. Methane is a hydrogen source of increasing importance.

Isotopes of Hydrogen

Hydrogen has three main isotopes: protium (^1H), deuterium (^2H) and tritium (^3H). These isotopes form naturally in nature. Protium and deuterium are stable. Tritium is radioactive and has a half-life of about 12 years. Scientists have created four other hydrogen isotopes (^4H to ^7H), but these isotopes are very unstable and do not exist naturally.

The main isotopes of hydrogen are unique because they are the only isotopes that have a name. These names are still in use today. Deuterium and tritium sometimes get their own symbols, D and T. However, the International Union of Pure and Applied Chemistry does not like these names very much, even though they are often used. There are other isotopes that had their own names when scientists studied radioactivity.

Protium

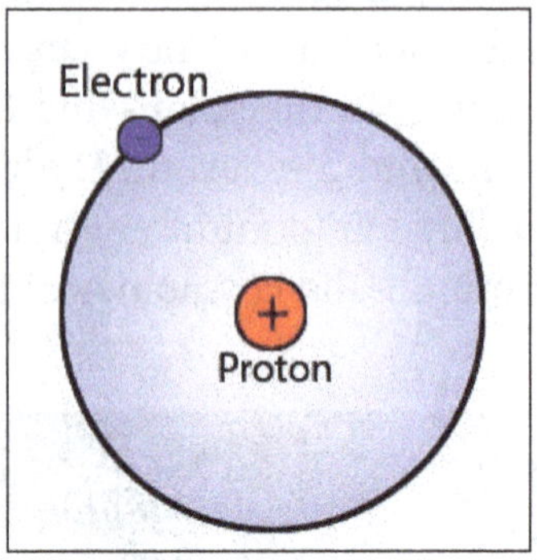

Protium, the most common isotope of hydrogen. It is special because it is the only isotope that has no neutron.

Protium is the most common isotope of hydrogen. It makes up more than 99.98% of all the hydrogen in the universe. It is named protium because its nucleus only has one proton. Protium has an atomic mass of 1.00782504(7) u. The symbol for protium is ^1H.

The proton of protium never decayed. So, scientists believe that protium is a stable isotope. New theories of particle physics predict that a proton can decay, but this decay is very slow. Proton is said to have a half-life of 10^{36} years. If proton decay is true, then all other nuclei that are said to be stable are actually only observationally stable, i.e. they look like they are stable. Recent experiments have shown that if proton decay does occur, it would have a half-life of 6.6×10^{33} years.

Deuterium

Deuterium (chemical symbol D or ^2H) is a stable isotope of hydrogen, found in extremely small amounts in nature. The nucleus of deuterium, called a deuteron, contains one proton and one neutron, whereas the far more common hydrogen nucleus contains just one proton and no neutrons. Consequently, each atom of deuterium has roughly twice the mass of an ordinary hydrogen atom, and deuterium is also called heavy hydrogen. Water in which ordinary hydrogen atoms are replaced by deuterium atoms is known as heavy water.

Scientists have developed a variety of applications for deuterium and its compounds. For example, deuterium serves as a nonradioactive isotopic tracer to study chemical reactions and metabolic pathways. In addition, it is useful for the study of macromolecules by neutron scattering. Deuterated solvents (such as heavy water) are routinely used in nuclear magnetic resonance (NMR) spectroscopy, because these solvents do not interfere with the NMR spectra of the compounds being studied. Deuterated compounds are also useful for femtosecond infrared spectroscopy. By measuring small variations in the natural abundance of deuterium, as well as variations of the stable heavy oxygen isotopes ^{17}O and ^{18}O, researchers can trace the geographic origins of Earth's waters. Deuterium is also a fuel for nuclear fusion reactions, which could someday be harnessed for commercial-scale power generation.

Natural Abundance

Deuterium occurs in trace amounts naturally as deuterium gas, written as $^2H_2$ or D_2. However, most deuterium atoms in the universe are bonded with typical ^1H atoms to form the gas called hydrogen deuteride (HD or $^1H^2H$).

The abundance of deuterium in the oceans of Earth is approximately one atom in 6500 hydrogen atoms (about 154 parts per million (ppm)). Deuterium thus accounts for approximately 0.015 percent (on a weight basis, 0.030 percent) of all naturally occurring hydrogen atoms in the oceans on Earth; the abundance changes slightly from one kind of natural water to another. The abundance of deuterium on Jupiter is about 6 atoms in 10,000 (0.06 percent atom basis). There is little deuterium in the interior of the Sun, because thermonuclear reactions destroy it. However, it persists in the outer solar atmosphere at roughly the same concentration as in Jupiter.

The existence of deuterium on Earth, elsewhere in the Solar System (as confirmed by planetary probes), and in stars (as indicated by their spectra), is an important piece of information in cosmology. Stellar fusion destroys deuterium, and there are no known natural processes (such as cluster decay), other than the Big Bang nucleosynthesis process, which might have produced deuterium at anything close to its observed natural abundance. This abundance seems to be a very similar fraction of hydrogen, wherever hydrogen is found. Thus, the existence of deuterium at its present abundance is one of the arguments in favor of the Big Bang theory over the steady state theory of the universe. It is estimated that the abundances of deuterium have not changed significantly since their production more than 14 billion years ago.

The world's leading "producer" of deuterium (technically, enricher or concentrator of deuterium) was Canada, until 1997, when the last plant was shut down. Canada uses heavy water as a neutron moderator for the operation of its CANDU reactors. Currently, India is probably the world's largest concentrator of heavy water, also used in nuclear power reactors.

Properties

The color, odor, and various chemical properties of deuterium are similar to those of protium. However, some of the physical properties of deuterium and its compounds differ from those of their ordinary hydrogen analogs. For example, the melting point of deuterium is -426 °F (-254 °C), whereas that of protium is -434 °F (-259 °C); and the boiling point of deuterium is -417 °F (-249 °C), whereas that of protium is -423 °F (-253 °C).

D_2O is more viscous than H_2O. Also, chemical bonds involving deuterium and tritium are somewhat stronger than the corresponding bonds in light hydrogen, and these differences lead to significant changes in biochemical reactions. The differences in bond energy and length for compounds of heavy hydrogen are greater than the isotopic differences for other elements.

Deuterium atoms can replace normal hydrogen atoms in water molecules to form heavy water (D_2O), which is about 10.6 percent denser than normal water. Consequently, ice made from heavy water sinks in ordinary water. Also, heavy water is slightly toxic in eukaryotic organisms, with 25 percent substitution of the body water causing cell division problems and sterility, and 50 percent substitution causing death by cytotoxic syndrome (bone marrow failure and gastrointestinal lining failure). Prokaryotic organisms, however, can survive and grow in pure heavy water (though they grow more slowly).

It appears that consumption of heavy water does not pose a health threat to humans unless very large quantities (in excess of 10 liters) were consumed over many days. Small doses of heavy water (a few grams in humans, containing an amount of deuterium comparable to that normally present in the body) are routinely used as harmless metabolic tracers in humans and animals.

Data of Deuterium

- Density: 0.180 kg/m³ at standard conditions for temperature and pressure (STP) (0 °C, 101.325 kPa).
- Atomic weight: 2.01355321270.
- Mean abundance in ocean water about 0.0156 percent of H atoms = 1/6400 H atoms.

Data at approximately 18 K for D_2 (triple point):

- Density:
    - Liquid: 162.4 kg/m³.
    - Gas: 0.452 kg/m³.
- Viscosity: 12.6 μPa·s at 300 Kelvin (gas phase).
- Specific heat capacity at constant pressure c_p:
    - Solid: 2950 J/(kg·K).
    - Gas: 5200 J/(kg·K).

Spectroscopic Differences with common Hydrogen

The nuclear magnetic resonance (NMR) frequency of deuterium is significantly different from that of common light hydrogen. Infrared spectroscopy also easily differentiates many deuterated compounds, because there is a large difference in infrared absorption frequency between the vibration of a chemical bond involving deuterium versus a bond involving light hydrogen. The two stable isotopes of hydrogen can also be distinguished by mass spectrometry.

Applications

Deuterium and its compounds are useful for a variety of applications. For example, in chemistry and biochemistry, deuterium is used as a nonradioactive isotopic tracer in molecules to study chemical reactions and metabolic pathways. Chemically, deuterium behaves much like ordinary hydrogen, but it can be distinguished from ordinary hydrogen by its mass, using mass spectrometry or infrared spectrometry.

Neutron scattering techniques particularly benefit from the availability of deuterated samples. The H and D cross-sections are very distinct and different in sign, which allows contrast variation in such experiments. Further, a nuisance problem of ordinary hydrogen is its large incoherent neutron cross-section, which is nil for D and delivers much clearer signals in deuterated samples. Hydrogen occurs in practically all organic chemicals and biochemicals, but it cannot be seen by X-ray diffraction methods. Hydrogen can be seen by neutron diffraction and scattering, which makes neutron scattering, together with a modern deuteration facility, indispensable for many studies of macromolecules in biology and other fields.

Deuterated solvents (including heavy water and compounds like deuterated chloroform, $CDCl_3$) are used in hydrogen nuclear magnetic resonance spectroscopy (proton NMR). NMR ordinarily requires compounds of interest to be analyzed after being dissolved in solution. Because deuterium's nuclear spin properties differ from those of light hydrogen present in organic molecules, NMR spectra of hydrogen/protium are clearly distinguishable from those of deuterium. In practice, deuterium is not "seen" by an NMR instrument tuned to light-hydrogen. A deuterated solvent therefore allows the light-hydrogen spectrum of the compound of interest to be measured, without interference from the solvent.

Deuterated compounds can also be used for femtosecond infrared spectroscopy, because the mass difference (compared to protium-containing compounds) drastically affects the frequency of molecular vibrations. Deuterium-carbon bond vibrations are found in locations free of other signals.

Deuterium is a fuel for nuclear fusion reactions, especially in combination with tritium, because of the high reaction rate (or nuclear cross section) and high energy yield of the D-T reaction. Unlike protium, deuterium undergoes fusion solely by the strong interaction, making its use for commercial power plausible.

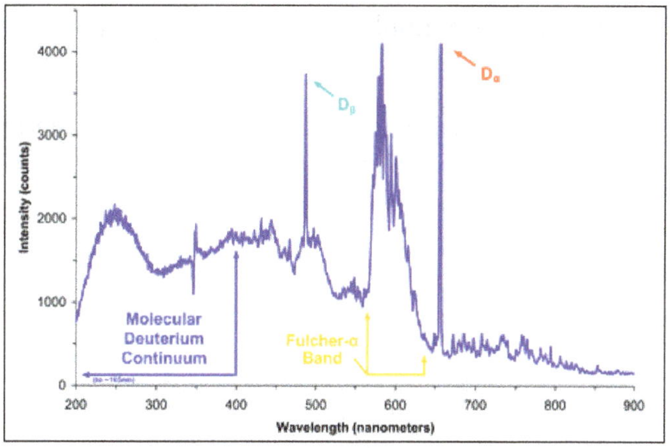

Emission spectrum of an ultraviolet deuterium arc lamp.

Measurements of small variations in the natural abundance of deuterium, along with those of the stable heavy oxygen isotopes ^{17}O and ^{18}O, are of importance in hydrology, to trace the geographic origins of Earth's waters. The heavy isotopes of hydrogen and oxygen in rainwater (also called meteoric water) are enriched as a function of the environmental temperature of the region in which precipitation occurs (and thus enrichment is related to mean latitude). The relative enrichment of heavy isotopes in rainwater (as referenced to mean ocean water), when plotted against temperature, falls predictably along a line called the global meteoric water line (GMWL). This plot allows samples of precipitation-originated water to be identified along with general information about the climate in which it originated. Evaporative and other processes in bodies of water, and also groundwater processes, also differentially alter the ratios of heavy hydrogen and oxygen isotopes in fresh and salt waters, in characteristic and often regionally distinctive ways.

The proton and neutron making up the deuterium nucleus can be dissociated through neutral current interactions with neutrinos. The cross-section for this interaction is comparatively large, and deuterium was successfully used as a neutrino target in the Sudbury Neutrino Observatory experiment.

Nuclear Physics of the Deuterium Nucleus

Deuterium is one of only four stable nuclides (^{2}H, ^{6}Li, ^{10}B, ^{14}N) with an odd number of protons and odd number of neutrons. Also the long-lived radioactive nuclides ^{40}K, ^{50}V, ^{138}La, ^{180m}Ta occur naturally and have the odd-odd combination of protons and neutrons. Most odd-odd nuclei are unstable with respect to beta decay, because the decay products are even-even, and are therefore more strongly bound, due to nuclear pairing effects. Deuterium, however, benefits from having its proton and neutron coupled to a spin-1 state, which leads to stronger nuclear attraction between the particles.

Deuterium Nucleus as an Isospin Singlet

Given that the proton and neutron are similar to each other in mass and nuclear properties, they are sometimes considered two symmetric types of the same object, a nucleon. Although only the proton has an electric charge, this is often neglected for nuclear interactions, because of the weakness of the electromagnetic interaction relative to the strong nuclear interaction. The symmetry relating the proton and neutron is known as isospin, denoted by the symbol τ.

The symmetry of isospin is SU(2), like that of ordinary spin, so the two are mutually analogous. The proton and neutron form an isospin doublet, with a "down" state $\downarrow$ being a neutron, and an "up" state $\uparrow$ being a proton.

A pair of nucleons can either be in an antisymmetric state of isospin called singlet, or in a symmetric state called triplet. In terms of the "down" state and "up" state, the singlet may be written as:

$$\frac{1}{\sqrt{2}}\left(|\uparrow\downarrow\rangle - |\downarrow\uparrow\rangle\right)$$

This is a nucleus with one proton and one neutron, that is, a stable deuterium nucleus.

The triplet may be written as,

$$\begin{pmatrix} \uparrow\uparrow \\ \frac{1}{\sqrt{2}}(\uparrow\downarrow + \downarrow\uparrow) \\ \downarrow\downarrow \end{pmatrix}$$

The triplet state consists of three types of nuclei: a highly excited state of the deuterium nucleus; a nucleus with two protons; and a nucleus with two neutrons. These three types of nuclei are not stable or nearly stable.

Approximated Wave Function of the Deuteron

The total wave function of both the proton and neutron must be antisymmetric, because they are both fermions. Apart from their isospin, the two nucleons also have spin and spatial distributions of their wave function. The latter is symmetric if the deuteron is symmetric under parity (i.e. have an "even" or "positive" parity), and antisymmetric if the deuteron is antisymmetric under parity (i.e. have an "odd" or "negative" parity). The parity is fully determined by the total orbital angular momentum of the two nucleons: if it is even, then the parity is even (positive); and if it is odd, the parity is odd (negative).

The deuteron, being an isospin singlet, is antisymmetric under nucleons exchange due to isospin, and therefore must be symmetric under the double exchange of their spin and location. Therefore it can be in either of the following two states:

- Symmetric spin and symmetric under parity: In this case, the exchange of the two nucleons will multiply the deuterium wave function by (-1) from isospin exchange, (+1) from spin exchange, and (+1) from parity (location exchange), for a total of (-1) as needed for antisymmetry.
- Antisymmetric spin and antisymmetric under parity: In this case, the exchange of the two nucleons will multiply the deuterium wave function by (-1) from isospin exchange, (-1) from spin exchange, and (-1) from parity (location exchange), again for a total of (-1) as needed for antisymmetry.

In the first case, the deuteron is a spin triplet, so that its total spin s is 1. It also has an even parity and therefore even orbital angular momentum l. The lower its orbital angular momentum, the lower its energy. Therefore the lowest possible energy state has $s = 1, l = 0$.

In the second case, the deuteron is a spin singlet, so that its total spin s is 0. It also has an odd parity and therefore odd orbital angular momentum l. Therefore the lowest possible energy state has $s = 0, l = 1$.

Because the nuclear attraction is stronger when $s = 1$, the deuterium ground state has the values $s = 1, l = 0$.

The same considerations lead to the possible states of an isospin triplet having $s = 0$, l = even; or $s = 1$, l = odd. Thus the state of lowest energy has $s = 1, l = 1$, higher than that of the isospin singlet.

The analysis just given is only approximate, because isospin is not an exact symmetry and more importantly because the strong nuclear interaction between the two nucleons is related to angular momentum in a way that mixes different s and l states. That is, s and l are not constant in time, and over time, a state such as $s = 1, l = 0$ may become a state of $s = 1, l = 2$. Parity is still constant in time, so these do not mix with odd l states (such as $s = 0, l = 1$). Therefore the quantum state of the deuterium is a superposition (a linear combination) of the $s = 1, l = 0$ state and the $s = 1, l = 2$ state, even though the first component is much bigger. Since the total angular momentum j is also a good quantum number (it is a constant in time), both components must have the same j, and therefore $j = 1$. This is the total spin of the deuterium nucleus.

To summarize, the deuterium nucleus is antisymmetric in terms of isospin, and has spin 1 and even (+1) parity. The relative angular momentum of its nucleons l is not well defined, and the deuterium nucleus is a superposition of mostly $l = 0$ with some $l = 2$.

Magnetic and Electric Multipoles

To find the theoretical value of the deuteron magnetic dipole moment μ, one uses the formula for nuclear magnetic moment:

$$\mu = \frac{1}{(j+1)} \left\langle (l,s), j, m_j = j \middle| \vec{\mu} \cdot \vec{j} \middle| (l,s), j, m_j = j \right\rangle$$

with,

$$\vec{\mu} = g^{(l)}\vec{l} + g^{(s)}\vec{s}$$

$g^{(l)}$ and $g^{(s)}$ are g-factors of the nucleons.

Since the proton and neutron have different values for $g^{(l)}$ and $g^{(s)}$, one must separate their contributions. Each gets half of the deuterium orbital angular momentum $\vec{l}$ and spin $\vec{s}$. One arrives at,

$$\mu = \frac{1}{(j+1)}\left\langle (l,s), j, m_j = j \left| \left(\frac{1}{2}\vec{l}\, g^{(l)}{}_p + \frac{1}{2}\vec{s}(g^{(s)}{}_p + g^{(s)}{}_n) \right) \cdot \vec{j} \right| (l,s), j, m_j = j \right\rangle$$

where subscripts p and n stand for the proton and neutron, and $g^{(l)}{}_n = 0$.

By using the same identities as here and using the value $g^{(l)}{}_p = 1$ in nuclear magneton units, we arrive at the following result, in nuclear magneton units:

$$\mu = \frac{1}{4(j+1)}\left[g^{(s)}{}_p + g^{(s)}{}_n (j(j+1) - l(l+1) + s(s+1)) + (j(j+1) + l(l+1) - s(s+1)) \right]$$

For the $s = 1$, $l = 0$ state, $j = 1$ and we get, in nuclear magneton units,

$$\mu = \frac{1}{2}(g^{(s)}{}_p + g^{(s)}{}_n) = 0.879$$

For the $s = 1$, $l = 2$ state with $j = 1$ we get, in nuclear magneton units,

$$\mu = -\frac{1}{4}(g^{(s)}{}_p + g^{(s)}{}_n) + \frac{3}{4} = 0.310$$

The measured value of the deuterium magnetic dipole moment, in nuclear magneton units, is 0.857. This suggests that the state of the deuterium is indeed only approximately $s = 1$, $l = 0$ state, and is actually a linear combination of (mostly) this state with $s = 1$, $l = 2$ state. The electric dipole is zero as usual.

The measured electric quadrupole of the deuterium is 0.2859 e fm^2, where e is the proton electric charge and fm is fermi. While the order of magnitude is reasonable, since the deuterium radius is of order of 1 fermi and its electric charge is e, the above model does not suffice for its computation. More specifically, the electric quadropole does not get a contribution from the $l = 0$ state (which is the dominant one) and does get a contribution from a term mixing the $l = 0$ and the $l = 2$ states, because the electric quadrupole operator does not commute with angular momentum. The latter contribution is dominant in the absence of a pure $l = 0$ contribution, but cannot be calculated without knowing the exact spatial form of the nucleons wave function inside the deuterium.

Higher magnetic and electric multipole moments cannot be calculated by the above model, for similar reasons.

Antideuteron and Antideuterium

An antideuteron is the antiparticle of the nucleus of deuterium, consisting of an antiproton and an antineutron. The antideuteron was first produced in 1965 at the Proton Synchrotron at CERN and the Alternating Gradient Synchrotron at Brookhaven National Laboratory. A complete atom, with a positron orbiting the nucleus, would be called antideuterium, but antideuterium has not been created as of 2005. The symbol for antideuterium is the same as for deuterium, except with a bar over it.

Pycnodeuterium

Deuterium atoms can be absorbed into a palladium (Pd) lattice. They are effectively solidified as an ultrahigh-density lump of deuterium lump, called *pycnodeuterium,* inside each octahedral space within the unit cell of the Pd host lattice. The authors believe this can be used as a nuclear fuel to perform cold fusion. Although this mechanism results in high concentrations of deuterium, the possibility of cold fusion has not been generally accepted within the scientific community.

Tritium

Tritium (chemical symbol T or ^3H) is a radioactive isotope of hydrogen. The nucleus of tritium (sometimes called a triton) contains one proton and two neutrons, whereas the nucleus of protium (the most abundant hydrogen isotope) contains one proton and no neutrons. Tritium emits low-energy beta radiation that cannot penetrate human skin, so this isotope is dangerous only if inhaled or ingested.

The properties of tritium make it useful for various applications. For instance, it is used in some self-illuminating watches, compasses, key chains, and gun sights for firearms. It is occasionally used as a radioactive label in molecules to trace their reactions and biochemical pathways. It is widely used in nuclear weapons for boosting a fission bomb or the fission primary of a thermonuclear weapon. Tritium is also an important fuel for controlled nuclear fusion reactions.

Radioactive Decay

The radioactive decay of tritium may be classified as beta decay (β decay). The nuclear reaction may be written as follows:

$$^{3}_{1}\mathrm{T} \rightarrow {}^{3}_{2}\mathrm{He} + \mathrm{e}^{-} + \bar{\nu}_{\mathrm{e}}$$

Each such reaction produces helium-3, an electron, and a nearly undetectable electron antineutrino, along with about 18.6 keV of energy. The electron has an average kinetic energy of 5.7 keV, while the remaining energy is carried off by the electron antineutrino.

Although experiments have indicated somewhat different values for the half-life of tritium, the U.S. National Institute of Standards and Technology (NIST) recommends a value of 4500±8 days (approximately 12.32 years).

The low-energy beta radiation from tritium cannot penetrate human skin, so tritium is dangerous only if inhaled or ingested. This low energy makes it difficult to detect tritium-labeled compounds, except by the technique of liquid scintillation counting.

Natural and Artificial Production

In nature, tritium is produced by the interaction of cosmic rays with atmospheric gases. In the most significant reaction for natural tritium production, a fast neutron (of energy greater than 4 MeV) interacts with atmospheric nitrogen-14 to produce carbon-12 and tritium, as follows:

$$^{14}_{7}\mathrm{N} + \mathrm{n} \rightarrow {}^{12}_{6}\mathrm{C} + {}^{3}_{1}\mathrm{T}$$

Because of tritium's relatively short half-life, however, tritium produced in this manner does not accumulate over geological timescales, and its natural abundance is negligible.

In nuclear reactors, tritium can be produced by the neutron activation of lithium-6, using neutrons of any energy. The reaction is exothermic, yielding 4.8 MeV of energy, which is more than one-quarter of the energy that can be produced by the fusion of a triton with a deuteron.

$$^{6}_{3}\mathrm{Li} + \mathrm{n} \rightarrow {}^{4}_{2}\mathrm{He}\ (2.05\,\mathrm{MeV}) + {}^{3}_{1}\mathrm{T}\ (2.75\,\mathrm{MeV})$$

High-energy neutrons can also produce tritium from lithium-7 in an endothermic reaction, consuming 2.466 MeV of energy. This reaction was discovered when the 1954 Castle Bravo nuclear test produced an unexpectedly high yield.

$$^{7}_{3}\mathrm{Li} + \mathrm{n} \rightarrow {}^{4}_{2}\mathrm{He} + {}^{3}_{1}\mathrm{T} + \mathrm{n}$$

When high-energy neutrons irradiate boron-10, tritium is occasionally produced. The more common result of boron-10 neutron capture is ^{7}Li and a single alpha particle.

$$^{10}_{5}\mathrm{B} + \mathrm{n} \rightarrow 2\,{}^{4}_{2}\mathrm{He} + {}^{3}_{1}\mathrm{T}$$

Reactions requiring high neutron energies are not attractive production methods.

Helium-3, produced during the beta decay of tritium, has a very large cross section for the (n, p) reaction with thermal neutrons. It is rapidly converted back to tritium in a nuclear reactor, as follows:

$$^{3}_{2}\mathrm{He} + \mathrm{n} \rightarrow {}_{1}\mathrm{H} + {}^{3}_{1}\mathrm{T}$$

Tritium is occasionally a direct product of nuclear fission, with a yield of about 0.01 percent (one per 10,000 fissions). This means that tritium release or recovery needs to be considered in nuclear reprocessing even in ordinary spent nuclear fuel where tritium production was not a goal.

Tritium is also produced in heavy water-moderated reactors when deuterium captures a neutron. This reaction has a very small cross section (which is why heavy water is such a good neutron moderator), and relatively little tritium is produced. Nevertheless, cleaning tritium from the moderator may be desirable after several years to reduce the risk of escape to the environment. Ontario Power Generation's Tritium Removal Facility can process up to 2.5 thousand metric tons (2,500 Mg) of heavy water a year, producing about 2.5 kg of tritium.

According to the 1996 report of the Institute for Energy and Environmental Research (IEER) about the United States Department of Energy, only 225 kg of tritium has been produced in the U.S. since 1955. Given that tritium is continuously decaying into helium-3, the stockpile was approximately 75 kg at the time of the report.

Tritium for American nuclear weapons was produced in special heavy water reactors at the Savannah River Site until their shutdown in 1988. With the Strategic Arms Reduction Treaty after the end of the Cold War, existing supplies were sufficient for the new, smaller number of nuclear weapons for some time. Production was resumed with irradiation of lithium-containing rods (replacing the usual boron-containing control rods) at the commercial Watts Bar Nuclear Generating Station in 2003-2005, followed by extraction of tritium from the rods at the new Tritium Extraction Facility at SRS starting in November 2006.

Properties

Tritium has an atomic mass of 3.0160492. It is a gas (T_2 or $^3H_2$) at standard temperature and pressure. It combines with oxygen to form a liquid called tritiated water, T_{2O}, or partially tritiated water, THO.

Tritium figures prominently in studies of nuclear fusion because of its favorable reaction cross section and the large amount of energy (17.6 MeV) produced through its reaction with deuterium:

$$^3_1\mathrm{T} + ^2_1\mathrm{D} \rightarrow ^4_2\mathrm{He} + \mathrm{n}$$

All atomic nuclei, being composed of protons and neutrons, repel one another because of their positive charge. However, if the atoms have a high enough temperature and pressure (for example, in the core of the Sun), then their random motions can overcome such electrical repulsion (called the Coulomb force), and they can come close enough for the strong nuclear force to take effect, fusing them into heavier atoms.

The tritium nucleus, containing one proton and two neutrons, has the same charge as the nucleus of ordinary hydrogen, and it experiences the same electrostatic repulsive force when brought close to another atomic nucleus. However, the neutrons in the tritium nucleus increase the attractive strong nuclear force when brought close enough to another atomic nucleus. As a result, tritium can more easily fuse with other light atoms, compared with the ability of ordinary hydrogen to do so.

The same is true, albeit to a lesser extent, of deuterium. This is why brown dwarfs (so-called failed stars) cannot burn hydrogen, but they do indeed burn deuterium.

Like hydrogen, tritium is difficult to confine. Rubber, plastic, and some kinds of steel are all somewhat permeable. This has raised concerns that if tritium is used in quantity, in particular for fusion reactors, it may contribute to radioactive contamination, although it short half-life should prevent significant long-term accumulation in the atmosphere.

Atmospheric nuclear testing (prior to the Partial Test Ban Treaty) proved unexpectedly useful to oceanographers, as the sharp spike in surface tritium levels could be used over the years to measure the rate of mixing of the lower and upper ocean levels.

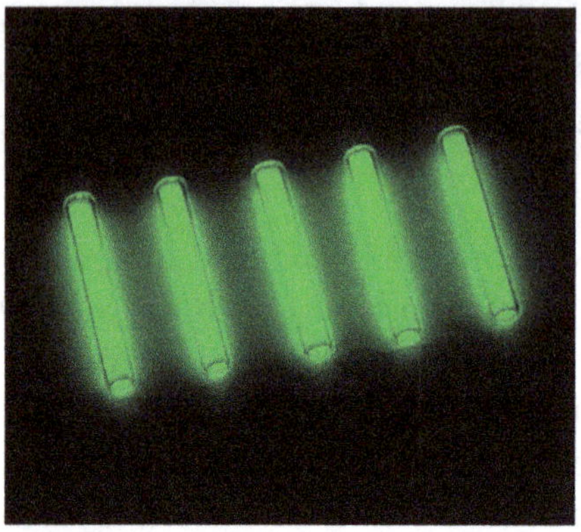

This "gaseous tritium light source," about 1.5 years old, is a glass vial filled with tritium gas. The inner surface of the vial is coated with a phosphor, which glows as it is continually struck by electrons emitted by the tritium.

Applications

Self-powered Lighting

The electrons emitted by small amounts of tritium can cause phosphors to glow. This phenomenon is employed in self-powered lighting devices called trasers, which are now used in watches and exit signs. It is also used in certain countries to make glowing key chains and compasses. In recent years, the same process has been used to make self-illuminating gun sights for firearms, especially semi-automatic handguns. The tritium takes the place of radium, which can cause bone cancer. Such uses of radium have been banned in most countries for decades.

According to the aforementioned IEER report, the commercial demand for tritium is about 400 grams per year.

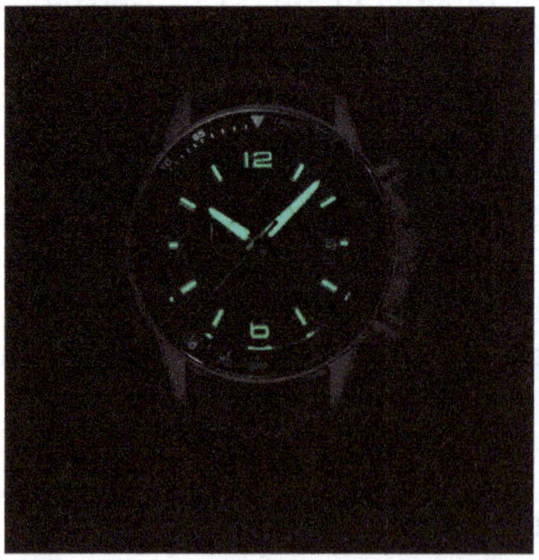

A watch face illuminated with tritium.

Analytical Chemistry

Tritium is sometimes used as a radioactive label in molecules to trace their reactions and pathways. Given that hydrogen appears in almost all organic chemicals, it is easy to find a place to put tritium on the molecule under investigation. The disadvantage of using tritium is that it produces a comparatively weak signal.

Nuclear Weapons

Tritium is widely used in nuclear weapons for boosting a fission bomb or the fission primary of a thermonuclear weapon. Before detonation, a small quantity (few grams) of tritium-deuterium gas is injected into the hollow "pit" of fissile plutonium or uranium. The early stages of the fission chain reaction supply enough heat and compression to start DT fusion. Thereafter, both fission and fusion proceed in parallel, the fission assisting the fusion by continued heating and compression, and the fusion assisting the fission with highly energetic (14.1 MeV) neutrons. As the fission fuel becomes depleted and also explodes outward, it falls below the density needed to stay critical by itself, but the fusion neutrons make the fission process progress faster and continue longer than it would without boosting. Increased yield (compared to the use of fission fuel without boosting) comes overwhelmingly from the increase in fission; the energy released by the fusion itself is much smaller because the amount of fusion fuel is much smaller.

Besides providing increased yield, tritium offers the possibility of variable yield, by varying the amount of fusion fuel. Perhaps even more significantly, tritium allows the weapon (or primary of a weapon) to have a smaller amount of fissile material (eliminating the risk of predetonation by nearby nuclear explosions) and more relaxed requirements for implosion, allowing a smaller implosion system.

Because tritium in the warhead is continuously decaying, it is necessary to replenish it periodically. The estimated quantity needed is four grams per warhead. To maintain constant inventory, 0.22 grams per warhead per year needs to be produced.

As tritium quickly decays and is difficult to contain, the much larger secondary charge of a thermonuclear weapon instead uses lithium deuteride (LiD) as its fusion fuel. During detonation, neutrons split lithium-6 into helium-4 and tritium; the tritium then fuses with deuterium, producing more neutrons. As this process requires a higher temperature for ignition, and produces fewer and less energetic neutrons (only D-D fusion and ^{7}Li splitting are net neutron producers), LiD is used only for secondaries, not for boosting.

Controlled Nuclear Fusion

Tritium is an important fuel for controlled nuclear fusion in both magnetic confinement and inertial confinement fusion reactor designs. The experimental fusion reactor ITER and the National Ignition Facility (NIF) will use Deuterium-Tritium (D-T) fuel. The D-T reaction is favored because it has the largest fusion cross section (~5 barns peak) and reaches this maximum cross section at the lowest energy (~65 keV center-of-mass) of any potential fusion fuel.

Bohr Model of Hydrogen Atom

In Bohr's model, the electron is pulled around the proton in a perfectly circular orbit by an attractive Coulomb force. The proton is approximately 1800 times more massive than the electron, so the proton moves very little in response to the force on the proton by the electron.

Bohr model of the hydrogen atom attempts to plug in certain gaps as suggested by Rutherford's model by including ideas from the newly developing Quantum hypothesis. According to Rutherford's model, an atom has a central nucleus and electron/s revolve around it like the sun-planet system. However, the fundamental difference between the two is that, while the planetary system is held in place by the gravitational force, the nucleus-electron system interacts by Coulomb's Law of Force. This is because the nucleus and electrons are charged particles. Also, an object moving in a circle undergoes constant acceleration due to the centripetal force.

Further, electromagnetic theory teaches us that an accelerating charged particle emits radiation in the form of electromagnetic waves. Therefore, the energy of such an electron should constantly decrease and the electron should collapse into the nucleus. This would make the atom unstable.

The classical electromagnetic theory also states that the frequency of the electromagnetic waves emitted by an accelerating electron is equal to the frequency of revolution. This would mean that, as the electron spirals inwards, it would emit electromagnetic waves of changing frequencies. In other words, it would emit a continuous spectrum. However, actual observation tells us that the electron emits a line spectrum.

Bohr Model Postulates

Bohr, in an attempt to understand the structure of an atom better, combined classical theory with the early quantum concepts and gave his theory in three postulates:

Postulate I: In a radical departure from the established principles of classical mechanics and electromagnetism, Bohr postulated that in an atom, electron/s could revolve in stable orbits without emitting radiant energy. Further, he stated that each atom can exist in certain stable states. Also, each state has a definite total energy. These are stationary states of the atom.

Postulate II: Bohr defined these stable orbits in his second postulate. According to this postulate:

- An electron revolves around the nucleus in orbits.
- The angular momentum of revolution is an integral multiple of h/2p – where hàPlanck's constant [h = 6.6 x 10^{-34} J-s].
- Hence, the angular momentum (L) of the orbiting electron is: L = nh/2p.

Postulate III: In this postulate, Bohr incorporated early quantum concepts into the atomic theory. According to this postulate, an electron can transitiòn from a non-radiating orbit to another of a lower energy level. In doing so, a photon is emitted whose energy is equal to the energy difference between the two states. Hence, the frequency of the emitted photon is:

hv = Ei – Ef

(Ei is the energy of the initial state and Ef is the energy of the final state. Also, Ei > Ef).

Radii of Bohr's Stationary Orbits

$$r_n = n^2\left(\frac{h^2\varepsilon_0}{\pi m Z e^2}\right)$$

where,

- n – integer,
- r_n – radius of the nth orbit,
- h – Planck's constant,
- ε_0 – Electric constant,
- m – Mass of the electron,
- Z – The Atomic number of the atom,
- e – Elementary charge.

Since ε_0, h, m, e, and p are constants and for a hydrogen atom, Z = 1, $r_n \propto n^2$.

Velocity of Electron in Bohr's Stationary Orbits:

$$v_n\left(\frac{Ze^2}{2h\varepsilon_0}\right)\left(\frac{1}{n}\right)$$

Since $_{\varepsilon 0}$, h, and e are constants and for a hydrogen atom, Z = 1, $r_n \propto (1/n)$.

Total Energy of Electron in Bohr's Stationary Orbits

$$E_n = \frac{me^4}{8\varepsilon_0{}^2h^2}\left(\frac{Z^2}{n^2}\right) \text{ or } E_n = -13.6\left(\frac{Z^2}{n^2}\right)ev$$

The negative sign means that the electron is bound to the nucleus.

Although these equations were derived under the assumption that electron orbits are circular, subsequent experiments conducted by Arnold Sommerfeld reaffirm the fact that the equations hold true even for elliptical orbits.

Energy Levels

When the electron is revolving in an orbit closest to the nucleus, the energy of the atom is the least or has the largest negative value. In other words, n = 1. For higher values of n, the energy is progressively larger.

The state of the atom wherein the electron is revolving in the orbit of smallest Bohr radius (a_0) is the 'Ground State'. In this state, the atom has the lowest energy. The energy in this state is:

$$E_1 = -13.6 \text{ eV}$$

Hence, the minimum energy required to free an electron from the ground state of an atom is 13.6 eV. This energy is the 'Ionization Energy' of the hydrogen atom. This value agrees with the experimental value of ionization energy too.

Now, a hydrogen atom is usually in 'Ground State' at room temperature. The atom might receive energy from processes like electron collision and acquire enough energy to raise the electron to higher energy states or orbits. This is an 'excited' state of the atom. Therefore, the energy required by the atom to excite an electron to the first excited state is:

$$E_2 - E_1 = -3.40 \text{ eV} - (-13.6) \text{ eV} = 10.2 \text{ eV}$$

Similarly, to excite the electron to the second excited state, the energy needed is:

$$E_3 - E_1 = -1.51 \text{ eV} - (-13.6) \text{ eV} = 12.09 \text{ eV}$$

Remember, that the electron can jump to a lower energy state by emitting a photon. Also, note that, as the excitation of the hydrogen atom increases, the minimum energy required to free the electron decreases.

Quantum Mechanical Model of Hydrogen Atom

The H atom is a bound state of a proton and an electron. The masses of the two particles are respectively:

$$m_p = 1.7 \times 10^{-27} \text{ kg},$$

$$m_e = 0.91 \times 10^{-30} \text{ kg}.$$

They have opposite charges, q and −q, with:

$$q = 1.6 \times 10^{-19} \text{ C}.$$

The ratio of the two masses is:

$$m_p / m_e = 1836.15267247(80);$$

it is known to 5 parts per billion.

The interaction between the two particles is due to electromagnetism; in a nonrelativistic formulation we can therefore model the H atom as a particle of reduced mass m:

$$\frac{1}{\mu} = \frac{1}{m_p} + \frac{1}{m_e} \approx \frac{1}{0.995 m_e},$$

in a Coulomb potential:

$$V(r) = \frac{q^2}{4\pi \epsilon_0 r} \equiv -\frac{e^2}{r}.$$

Given that the proton mass is much larger than the electron one, the reduced mass of the ep system is very close to the electron mass. The distance r that appears in the expression for the Coulomb potential is the distance between the electron and the proton. We can identify the origin of our reference frame with the position of the proton; the potential is clearly symmetric under rotations around the origin. The parameters that specify the physical system are the reduced mass μ and the electron charge e.

The H atom is an example of motion in a central force.

Stationary States

The Hamiltonian for the system is:

$$\hat{H} = -\frac{\hbar^2}{2\mu}\nabla^2 - \frac{e^2}{r}.$$

The time-independent Schrödinger equation,

$$\hat{\mathrm{H}}\psi(\underline{\mathrm{r}}) = \mathrm{E}\psi(\underline{\mathrm{r}}),$$

is solved as usual by separation of variables. Using for the solution $\psi(\underline{\mathrm{r}})$ the Ansatz:

$$\psi(\mathrm{r}, \theta, \varphi) = \frac{\chi\mathrm{nl}(\mathrm{r})}{\mathrm{r}} \mathrm{Y}_\ell^{\mathrm{m}}(\theta, \phi),$$

we obtain the radial one-dimensional equation:

$$\left[-\frac{\hbar^2}{2\mu}\frac{d^2}{dr^2} + \frac{\hbar^2\ell(\ell+1)}{2\mu r^2} - \frac{e^2}{r}\right]\chi_{nl}(r) = Enl\chi_{nl}(r).$$

Equation $\left[-\frac{\hbar^2}{2\mu}\frac{d^2}{dr^2} + \frac{\hbar^2\ell(\ell+1)}{2\mu r^2} - \frac{e^2}{r}\right]\chi_{nl}(r) = Enl\chi_{nl}(r)$.is the specific case of the Coulomb potential. The effective potential in this case is:

$$V(r) = -\frac{{}^2}{r} + \frac{\hbar^2 \ \ell(\ell \quad 1)}{r}.$$

The effective potential is sketched in figure below. The boundary condition for the radial function $\chi_{nl}(r)$ is:

$$\chi_{nl}(0) = 0.$$

Remember that, since we are looking for the bound states of the system, we are only interested in solutions with negative energy, i.e. $E_{nl} < 0$.

Equation $\left[-\frac{\hbar^2}{2\mu}\frac{d^2}{dr^2} + \frac{\hbar^2\ell(\ell+1)}{2\mu r^2} - \frac{e^2}{r}\right]\chi_{nl}(r) = Enl\chi_{nl}(r)$. suggests that the energy levels will depend on the total angular momentum ℓ.

It is useful to describe the solutions of the Schrödinger equation in terms of two physical quantities; a characteristic length,

$$a_0 = \frac{\hbar^2}{\mu e^2} \approx 0.52\,\text{Å},$$

and energy,

$$E_I = \frac{\mu e^4}{2\hbar^2} \approx 13.6eV.$$

Let us denote by E_{nl} the energy levels, and introduce the dimensionless variables,

$$\rho = r/a_0,\ \lambda_{nl} = \sqrt{-E_{nl}/E_1}.$$

Equation $\left[-\frac{\hbar^2}{2\mu}\frac{d^2}{dr^2}+\frac{\hbar^2\ell(\ell+1)}{2\mu r^2}-\frac{e^2}{r}\right]\chi_{nl}(r) = Enl\chi_{nl}(r)$.can be written as,

$$\left[-\frac{\hbar^2}{2\mu}\frac{1}{a_0^2}\frac{d^2}{d\rho^2}+\frac{\hbar^2\ell(\ell+1)}{2\mu a_0^2\rho^2}-\frac{e^2}{a_0\rho}-E_{nl}\right]\chi_{nl}(a_{0\rho}) = 0.$$

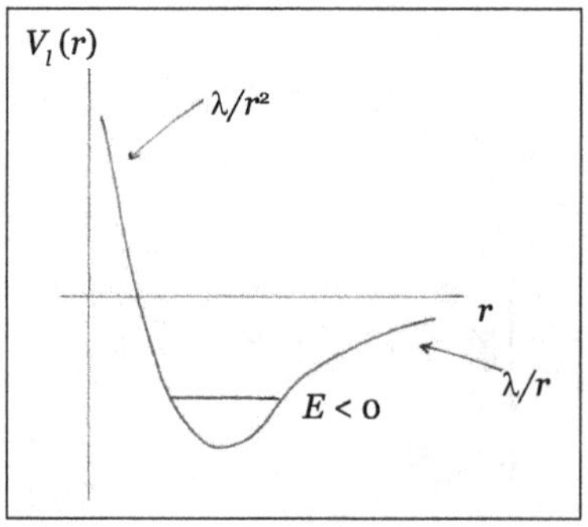

Effective potential for the one-dimensional radial Schrödinger equation for a system with total angular momentum *l*. We look for solutions of the time-independent. Schrödinger equation with negative energy *E* corresponding to bound states of the proton/electron system.

Introducing,

equation $\left[-\frac{\hbar^2}{2\mu}\frac{1}{a_0^2}\frac{d^2}{d\rho^2}+\frac{\hbar^2\ell(\ell+1)}{2\mu a_0^2\rho^2}-\frac{e^2}{a_0\rho}-E_{nl}\right]\chi_{nl}(a_{0\rho}) = 0$ can be rewritten as,

$$\left[\frac{d^2}{d\rho^2}-\frac{\ell(\ell+1)}{\rho^2}+\frac{2\mu e^2 a_0}{\hbar^2}\frac{1}{\rho}+\frac{2\mu a_0^2}{\hbar^2}En,\ell\right]u_{nl}(\rho) = 0,$$

and finally,

$$\left[\frac{d^2}{d\rho^2}-\frac{\ell(\ell+1)}{\rho^2}+\frac{2}{\rho}-\lambda_{nl}^2\right]u_{nl}(\rho) = 0;$$

We need to solve equation $\left[\frac{d^2}{d\rho^2}-\frac{\ell(\ell+1)}{\rho^2}+\frac{2}{\rho}-\lambda_{nl}^2\right]u_{nl}(\rho)=0$, in order to find the eigenvalues of the Hamiltonian E_{nl} and the corresponding radial wave functions $u_{nl}(r)$. Remember that the angular part of the wave functions is given by the spherical harmonics.

Solution of the Radial Equation

In this topic we shall discuss some technical details related to the solution of the radial equation $\left[\frac{d^2}{d\rho^2}-\frac{\ell(\ell+1)}{\rho^2}+\frac{2}{\rho}-\lambda_{nl}^2\right]u_{nl}(\rho)=0$. We wish to keep the mathematical details of the solution separated from the physical interpretation.

In order to shine a light on the form of the solution, we can start by considering its limiting behaviours as:

$$\rho\to 0 \text{ and } \rho\to\infty.$$

Let us first discuss the large-distance regime. As ρ increased, both the centrifugal and Coulomb potentials tend to zero and become unimportant in equation $\left[\frac{d^2}{d\rho^2}-\frac{\ell(\ell+1)}{\rho^2}+\frac{2}{\rho}-\lambda_{nl}^2\right]u_{nl}(\rho)=0$, which becomes:

$$\left[\frac{d^2}{d\rho^2}-\lambda_{nl}^2\right]u_{nl}(\rho)=0.$$

The solutions to this latter equation are simply,

$$u_{nl}(\rho)=\exp(\pm\lambda_{nl}\rho),$$

and the solution that grows exponentially must be discarded because it yields a non-normalizable wave function.

Of course we cannot completely neglect the potential terms, so we will look for a complete solution of the form:

$$u_{nl}(\rho)=e^{-\lambda nl\rho}\eta_{nl}(\rho).$$

Equation $\left[\frac{d^2}{d\rho^2}-\frac{\ell(\ell+1)}{\rho^2}+\frac{2}{\rho}-\lambda_{nl}^2\right]u_{nl}(\rho)=0$, becomes: $\eta_{nl}''-2\lambda_{nl}\eta_{nl}'+\left(-\frac{\ell(\ell+1)}{\rho^2}+\frac{2}{\rho}\right)\eta_{nl}=0$,

where the prime symbol ($'$) denotes the differentiation with respect to ρ. The boundary condition for u_{nl} translates into a boundary condition for u_{nl} namely $\eta_{nl}(0)=0$.

As $\rho \to 0$, we know that $u_{nl} \sim \rho^{\ell+1}$. Therefore we can look for a solution for u_{nl} expanded as a power series in ρ:

$$\eta_{nl}(\rho) = \rho^{\ell+1} \sum_{q=0}^{\infty} c_q \rho^q .$$

Inserting this Ansatz into equation $\eta''_{nl} - 2\lambda_{nl}\eta'_{nl} + \left(-\frac{\ell(\ell+1)}{\rho^2} + \frac{2}{\rho} \right) \eta_{nl} = 0$, we obtain:

$$\sum_q (q+l+1)(q+l)C_q \rho^{q+\ell-1} - 2\lambda_{nl}(q+\ell+1)c_q \rho^{q+\ell} + 2c_q \rho^{q+\ell} - \ell(\ell+1)c_q \rho^{q+\ell-1} = 0,$$

and hence:

$$\sum_q q(q+2\ell+1)c_q \rho^{q+\ell-1} - 2[\lambda_{nl}(q+\ell+1)-1]c_q \rho^{q+\ell} = 0.$$

Shifting the summation index, $q \to q-1$, in the second term of the sum above, we obtain:

$$\sum_q [q(q+2\ell+1)c_q - 2[\lambda_{nl}(q+\ell)-1]c_{q-1}]\rho^{q+\ell-1} = 0.$$

Since the last equality must hold for all values of ρ, we deduce:

$$q(q+2\ell+1)c_q - 2[\lambda_{nl}(q+l)-1]c_{q-1} = 0.$$

Equation $q(q+2\ell+1)c_q - 2[\lambda_{nl}(q+l)-1]c_{q-1} = 0$ is a recursion relation between the coefficients of the Taylor expansion of η_{nl} / ρ^{l+1}. It is crucial to note that for large q:

$$\frac{c_q}{c_{q-1}} \overset{q\to\infty}{\sim} \frac{2\lambda_{nl}}{q}, \text{ i.e. } c_q \sim \frac{(2\lambda_{nl})^q}{q!}.$$

The asymptotic behaviour for c_q would yield a solution:

$$\eta_{nl}(\rho) \sim \rho\infty^{\ell+1} e^{2\lambda_{nl}\rho},$$

which in turn yields a wave function u_{nl} that is not normalizable.

Therefore we must have $c_q = 0$ for some finite value of q, that we will denote $q = n_r > 0$. According to equation $q(q+2\ell+1)c_q - 2[\lambda_{nl}(q+l)-1]c_{q-1} = 0$, this can only happen if:

$$\lambda_{nl} = \frac{1}{n_r + \ell} \equiv \frac{1}{n}.$$

Then the expansion in equation $\eta_{nl}(\rho) = \rho^{\ell+1} \sum_{q=0}^{\infty} c_q \rho^q$ only contains a finite number of terms, i.e. it is simply a polynomial in ρ of finite order n_r.

We can see from equation $\lambda_{nl} = \frac{1}{n_r + \ell} \equiv \frac{1}{n}$ that this condition implies that the energy eigenvalues are quantized. Remember that λ_{nl} is related to the eigenvalues of the Hamiltonian via equation $\rho = r / a_0$, $\lambda_{nl} = \sqrt{-E_{nl} / E_1}$. Energy quantization is a consequence of having required the wave function to be normalizable.

Physical Interpretation

The computation shows that the eigenvalues of the Hamiltonian for the H atom are:

$$E_{nl} = \frac{-E_I}{(n_r + \ell)^2},$$

where ℓ is the angular momentum of the state, and $n_r > 0$ is an integer. We see that for the H atom the value of the energy does not depend on n_r and ℓ, but only on their sum $n = n_r + \ell$. The integer *n* is called the principal quantum number; its value characterizes the so-called electron shells.

We can rewrite equation $E_I = \frac{\mu e^4}{2\hbar^2} \approx 13.6 \text{ eV}$ as,

$$E_I = \frac{1}{2}\alpha^2 \mu c^2,$$

where α is the fine-structure constant,

$$\alpha = \frac{e^2}{hc} \simeq \frac{1}{137}.$$

Note that $\mu \simeq m_e$, and therefore μc^2 is the rest energy of the electron. Equation $E_I = \frac{1}{2}\alpha^2 \mu c^2$, shows that the typical scale of the energy levels in the H atom is 10^{-4} the rest energy of the electron. This justifies the nonrelativistic treatment of the H atom that we have used here. Clearly there are relativistic corrections; however they are small effect, typically of order α, and can be studied in perturbation theory.

Using equation $a_0 = \frac{\hbar^2}{\mu e^2} \approx 0.52\text{Å}$, we can express the energy levels as,

$$E_n = -\frac{e^2}{2n^2 a_0}, \quad n = n_r + l.$$

This the formula put forward by Bohr in 1913, before Quantum Mechanics was fully developed.

Equation $E_n = -\frac{e^2}{2n^2 a_0}$, $n = n_r + l$ shows clearly that for a given principal quantum number n, ℓ

can take the values $\ell = 0,1,\dots,n-1$ (corresponding respectively to $n_r = n, n-1,\dots,$). Since for each value of ℓ we have a $(2\ell+1)$ degeneracy, the total degeneracy of the level E_n is:

$$\sum_{\ell=0}^{n-1}(2\ell+1) = n(n-1)+n = n^2.$$

The polynomials w_{nl} are called the associated Laguerre polynomials. The full solution for the eigenfunctions of the energy is:

$$\Psi_{nlm}(\underline{r}) = R_{nl}(r)Y_\ell^m(\theta,\phi).$$

The first few radial functions are:

$$R_{1,0} = 2a_0^{-3/2}e^{-r/a_0},$$

$$R_{2,0} = \frac{1}{2\sqrt{2}}a_0^{-3/2}\left(2-\frac{r}{a_0}\right)e^{-r/(2a_0)},$$

$$R_{2,1} = \frac{1}{2\sqrt{6}}a_0^{-3/2}\frac{r}{a_0}e^{-r/(2a_0)}.$$

- The formula for the energy levels reproduces the Bohr spectrum, in agreement with the experimental data. The degeneracy of each level can only be obtained by the proper quantum-mechanical description that we have presented.
- The Bohr radius a_0 is the typical spatial extension of the ground state.
- The treatment described here can be applied to any hydrogen-like atom, i.e. an atom with an electron and a nucleus of charge Zq. Simply replace everywhere $e^2 \rightarrow Ze^2$
- It is interesting to compute the expectation value of the momentum. Since the typical size of the atom is a_0 we deduce from the Heisenberg uncertainty relation that $\sqrt{\langle p^2 \rangle} \sim \mu e^2 / h.$ Thus we obtain:

$$v \sim p/\mu \simeq e^2/\hbar = \alpha c \simeq \frac{1}{137}c << c.$$

We see a posteriori that motion of the electron is nonrelativistic, and hence our nonrelativistic description is accurate. Relativistic corrections are expected to be $O(v/c) \simeq O(\alpha)$.

The ionization energy is the energy needed to extract the electron from the bound state and is E_I This is known as the Rybderg energy.

The electron in the H atom can go from one shell to a lower one by emitting a photon. The series of transitions from principal number $n \geq 2$ to $n = 1$ is called the Lyman series 2. The transitions are names by Greek letters: the transition from $n = 2$ to $n = 1$ is called Lyman-α, from 3 to 1 Lyman-β, etc. Likewise the transitions from $n \geq 3$ to $n = 2$ form the Balmer series.

Hydrogen Spectrum

According to the quantum theory of Niels Bohr, the hydrogen atom can only exist in discrete energy states whose energies are given by:

$$E_n = -\left(\frac{me^4}{8\varepsilon_o^2 h^2}\right)\frac{1}{n^2}$$

where e and m are the electron charge and mass, respectively; ε_o is the dielectric permittivity of free space; h is Planck's constant; and n is the quantum number (n = 1, 2, 3, ...).

Figure below shows the energy level scheme for the hydrogen atom as calculated from equation:

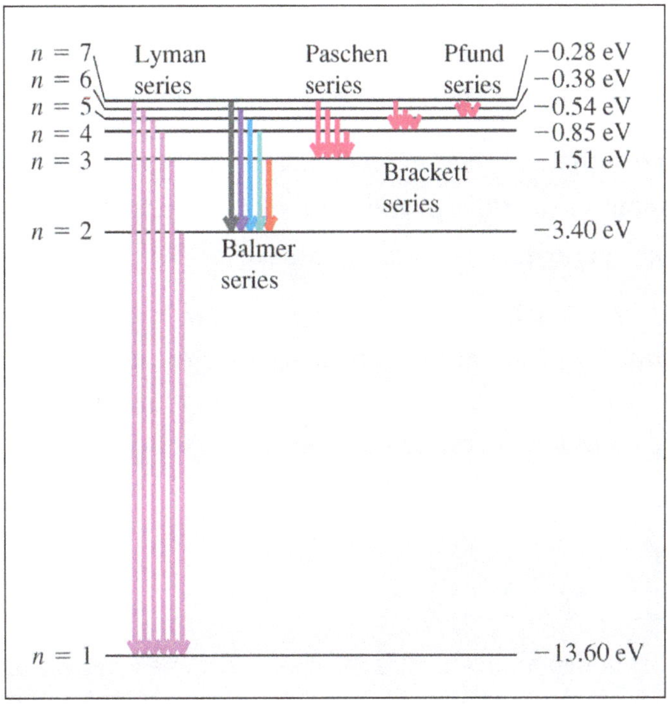

Under normal conditions, the hydrogen atom remains in its ground state (n = 1). It is possible to populate the various excited states (n > 1) by heating, electric discharge, irradiation, etc. The following decay from an excited state (n_i) to a lower-lying excited state or the ground state (n_f) is accompanied by the emission of electromagnetic radiation (a photon). The frequency of this radiation is determined by Bohr's second postulate:

$$\Delta E = hv$$

where ΔE is the energy difference between the two hydrogen atom states. Using equation $E_n = -\left(\frac{me^4}{8\varepsilon_o^2 h^2}\right)\frac{1}{n^2}$, for a transition from the initial energy state n_i to the final energy state n_f:

$$\Delta E = hv = E_i - E_f$$

$$hv = -\frac{me^4}{8\varepsilon_o^2 h^2}\left(\frac{1}{n_i^2}\right) - \left(-\frac{me^4}{8\varepsilon_o^2 h^2}\right)\left(\frac{1}{n_f^2}\right)$$

$$hv = \frac{me^4}{8\varepsilon_o^2 h^2}\left(\frac{1}{n_f^2} - \frac{1}{n_i^2}\right)$$

Now $v = c\lambda$, so:

$$\frac{1}{\lambda} = \frac{me^4}{8\varepsilon_o^2 h^3 c}\left(\frac{1}{n_f^2} - \frac{1}{n_i^2}\right)$$

$$\frac{1}{\lambda} = R\left(\frac{1}{n_f^2} - \frac{1}{n_i^2}\right)$$

where R = 1.097373 × 10^7 m^{-1} (the Rydberg constant).

All transitions which end at the same energy level, called final state (n_f) form a series. The various series are named in honour of their discoverers. $n_f = 1$ is the Lyman series, $n_f = 2$ is the Balmer series, $n_f = 3$ is the Paschen series, $n_f = 4$ is the Bracket series, and $n_f = 5$ is the Pfund series.

The first four transitions of the Balmer series of hydrogen can be observed as a line emission spectrum in the visible region. They are traditionally named H_α (red, $n_i = 3$), H_β (blue-green, $n_i = 4$), H_γ (violet, $n_i = 5$), and H_δ (violet, $n_i = 6$). The H_δ line is of very low intensity and so the lines originating in higher energy states are of even lower intensity.

Hydrogen-like Atoms

A hydrogen-like atom (or *hydrogenic* atom) is an atom with one electron. Except for the hydrogen atom itself (which is neutral) these atoms carry positive charge *e*(*Z*-1), where *Z* is the atomic number of the atom and *e* is the elementary charge. A better—but never used—name would therefore be hydrogen-like cations.

Because hydrogen-like atoms are two-particle systems with an interaction depending only on the distance between the two particles, their non-relativistic Schrödinger equation can be solved in analytic form. The solutions are one-electron functions and are referred to as hydrogen-like atomic orbitals. The orbitals of the different hydrogen-like atoms differ from one another in one respect only: they depend on the nuclear charge *eZ* (which appears in their radial part).

Hydrogen-like atoms *per se* do not play an important role in chemistry or physics. The interest in these atoms is mainly because their Schrödinger equation can be solved analytically, in exactly the same way as the Schrödinger equation of the hydrogen atom.

Quantum Numbers of Hydrogen-like Wave Functions

The non-relativistic wave functions (orbitals) of hydrogen-like atoms are known analytically and are labeled by three exact quantum numbers, conventionally designated n, ℓ, and m. These quantum numbers play an important role in atomic physics and chemistry, as they are useful labels for quantum mechanical states of more-electron atoms, too. Although the three quantum numbers are not exact for an atom with more than one electron, they are still approximately valid. Because the (for many-electron atoms) approximate quantum numbers n, ℓ, and m are the building bricks of the Aufbau principle (building-up principle)—the construction of the electronic configuration of atoms.

Eigenfunctions of Commuting Operators

Hydrogen-like atomic orbitals are eigenfunctions of a Hamiltonian H (energy operator) with eigenvalues proportional to $1/n^2$, where n is a positive integer, referred to as principal quantum number. Observe the somewhat unexpected fact that these eigenvalues do depend *solely* on n and not on other quantum numbers.

The hydrogen orbitals are usually chosen such that they are simultaneously eigenfunctions of H and l^2, the square of the one-electron angular momentum vector operator,

$$l \equiv -i\hbar(r \times \nabla) \equiv (l_x, l_y, l_z),$$

where $\hbar$ is Planck's constant divided by 2π, the symbol $\times$ stands for a cross product, ∇ is the gradient operator, and r is the vector pointing from the nucleus to the electron.

From quantum mechanics it is known that a necessary and sufficient condition for the existence of simultaneous eigenfunctions of H and l^2 is the commutation of the operators,

$$l^2 \equiv l_x^2 + l_y^2 + l_z^2 \text{ and } H.$$

These two operators indeed commute, because,

$$[H, l_\alpha] \equiv Hl_\alpha - l_\alpha H = 0, \text{ for } \alpha = x, y, z,$$

which is due to the spherical symmetry of H. From this follows the commutation, that is,

$$[H, l^2] = 0.$$

The squared operator l^2 has eigenvalues proportional to $\ell(\ell+1)$, where ℓ is a non-negative integer (the azimuthal quantum number, also known as the angular momentum quantum number).

Further, since l^2 commutes with the three angular momentum components l_x, l_y, and l_z,

$$[l^2, l_\alpha] = 0, \text{ for } \alpha = x, y, z,$$

it is possible to require an orbital to be an eigenfunction of any of the three components. It is conventional to choose l_z, which has an eigenvalue proportional to an integer usually denoted by m (the so-called magnetic quantum number).

Count the degenerate orbitals belonging to fixed n,

$$\sum_{\ell=0}^{n-1}\sum_{m=-\ell}^{\ell}1=\sum_{\ell=0}^{n-1}2\ell+1=n^2.$$

In other words, the degeneracy (maximum number of linearly independent eigenfunctions of same energy) of energy level n is equal to n^2. This is the dimension of the irreducible representations of the symmetry group of hydrogen-like atoms, which is SO(4), and not SO(3) as for many-electron atoms.

Quantum Numbers

A hydrogen-like atomic orbital $\psi_{n\ell m}$ is uniquely identified by the values of the principal quantum number n, the azimuthal quantum number ℓ, and the magnetic quantum number m. These three quantum numbers are natural numbers, their definitions and ranges are:

$$H\psi_{nlm}=\frac{E_h}{2n^2}\psi_{nlm}, \qquad n=1,\dots,\infty, \quad \text{(principal quantum number)}$$

$$l^2\psi_{nlm}=\hbar^2\ell(\ell+1)\psi_{nml}, \qquad \ell=0,\dots,n-1, \quad \text{(azimuthal quantum number)}$$

$$l_z\psi_{nlm}=\hbar m\psi_{nlm}, \qquad m=-\ell,\dots,\ell. \quad \text{(magnetic quantum number)}$$

Here E_h is the atomic unit of energy.

Indication of ℓ by Letters

It is very common to denote the orbitals of different angular momentum by different letters, 2*s*-, 3*p*-orbital, etc. For historical reasons ℓ = 0 orbitals are designated by *s* (sharp), ℓ = 1 by *p* (principal), ℓ = 2 by *d* (diffuse), and ℓ = 3 by *f* (fundamental). For higher ℓ orbitals the alphabet is followed, while *j* orbitals are omitted. Thus we get the following association between letters and ℓ quantum numbers,

s	*p*	*d*	*f*	*g*	*h*	*i*	*k*
0	1	2	3	4	5	6	7

For instance, hydrogenic *g*-orbitals (ℓ = 4) start at principal quantum number n = 5, so that we can speak of 5*g*-, 6*g*-, etc. orbitals, but a hydrogen-like 4*g*-orbital is not defined (i.e., does not appear as a solution of the hydrogen-like Schrödinger equation).

Spin

The set of orbital quantum numbers must be augmented by the two-valued spin quantum number $m_s = \pm½$ in application of the exclusion principle. This principle restricts the allowed values of the four quantum numbers in electron configurations of more-electron atoms: it is forbidden that two electrons have the same four quantum numbers. This is an important restriction in constructing atomic states by application of the Aufbau (building up) principle.

Schrödinger Equation

The atomic orbitals of hydrogen-like atoms are solutions of the time-independent Schrödinger equation in a potential given by Coulomb's law:

$$V(r) = -\frac{1}{4\pi \epsilon_0}\frac{Ze^2}{r}$$

where,

- ε_o is the permittivity of the vacuum,
- Z is the atomic number (charge of the nucleus in unit e; number of protons in the nucleus),
- e is the elementary charge (charge of an electron),
- r is the distance of the electron from the nucleus.

The Schrödinger equation is the following eigenvalue equation of the Hamiltonian (the quantity in large square brackets):

$$\left[-\frac{\hbar^2}{2\mu}\nabla^2 + V(r)\right]\psi(r) = E\psi(r),$$

where μ is the reduced mass of the system consisting of the electron and the nucleus. Because the electron mass is about 1836 times smaller than the mass of the lightest nucleus (the proton), the value of μ is very close to the mass of the electron m_e for all hydrogenic atoms. In the derivation below we will make the approximation $\mu = m_e$. Since m_e will appear explicitly in the formulas it will be easy to correct for this approximation if necessary.

The operator ∇^2 expressed in spherical polar coordinates, can be written as:

$$\hbar^2\nabla^2 = \frac{\hbar^2}{r}\frac{\partial^2}{\partial r^2}r - \frac{l^2}{r^2}.$$

The wave function is written as a product of functions in the spirit of the method of separation of variables:

$$\psi(r,\theta,\phi) = R(r)Y_{lm}(\theta,\phi),$$

where Y_{lm} are spherical harmonics, which are eigenfunctions of l^2 with eigenvalues $\hbar^2 l(l+1)$. Substituting this product, letting l^2 act on Y_{lm}, and dividing out Y_{lm}, we arrive at the following one-dimensional Schrödinger equation:

$$-\frac{\hbar^2}{2\mu}\left[\frac{1}{r}\frac{d^2}{dr^2}rR(r) - \frac{l(l+1)R(r)}{r^2}\right] + V(r)R(r) = ER(r).$$

Wave Function and Energy

In addition to *l* and *m*, there arises a third integer *n* > 0 from the boundary conditions imposed on *R*(r). The expression for the normalized wave function is:

$$\psi_{nlm} = R_{nl}(r)Y_{lm}(\theta,\phi),$$

where $Y_{lm}(\theta,\varphi)$ is a spherical harmonic. Below it will be derived that the radial function (normalized to unity) is,

$$R_{nl}(r) = \left(\frac{2Z}{na_\mu}\right)^{3/2} \left[\frac{(n-l-1)!}{2n[(n+l)!]}\right]^{1/2} e^{-Zr/na\mu} \left(\frac{2Zr}{na_\mu}\right)^{l} L_{n-l-1}^{2l+1}\left(\frac{2Zr}{na_\mu}\right).$$

Here:

- L_{n-l-1}^{2l+1} are the generalized Laguerre polynomials in the definition given here.
- $a_\mu = \dfrac{4\pi\varepsilon_0\hbar^2}{\mu e^2}$

Note that a_μ is approximately equal to a_0 (the Bohr radius). If the mass of the nucleus is infinite then $\mu = m_e$ and $a_\mu = a_0$.

The energy eigenvalue associated with ψ_{nlm} is:

$$E_n = -\frac{\mu}{2}\left(\frac{Ze^2}{4\pi\varepsilon_0\hbar n}\right)^2$$

As we pointed out above it depends only on *n*, not on *l* or *m*.

Derivation of Radial Function

As is shown above, we must solve the one-dimensional eigenvalue equation,

$$\left[-\frac{\hbar^2}{2m_e r}\frac{d^2}{dr^2}r + \frac{\hbar^2 l(l+1)}{2m_e r^2} + V(r)\right]R(r) = ER(r),$$

where we approximated μ by m_e. If the substitution *u(r)* = *rR(r)* is made, the radial equation becomes,

$$-\frac{\hbar^2}{2m_e}\frac{d^2u(r)}{dr^2} + V_{\text{eff}}(r)u(r) = Eu(r)$$

which is a Schrödinger equation for the function *u(r)* with an effective potential given by,

$$V_{\text{eff}}(r) = V(r) + \frac{\hbar^2 l(l+1)}{2m_e r^2}.$$

The correction to the potential *V(r)* is called the *centrifugal barrier term.*

In order to simplify the Schrödinger equation, we introduce the following constants that define the atomic unit of energy and length, respectively,

$$E_h = m_e \left(\frac{e^2}{4\pi\varepsilon_0 \hbar} \right)^2 \text{ and } a_0 = \frac{4\pi\varepsilon_0 \hbar^2}{m_e e^2}.$$

Substitute $y = Zr / a_0$ and $W = E / (Z^2 E_h)$ into the radial Schrödinger equation given above. This gives an equation in which all natural constants are hidden,

$$\left[-\frac{1}{2}\frac{d^2}{dy^2} + \frac{1}{2}\frac{l(l+1)}{y^2} - \frac{1}{y} \right] \mu_l = W_{ul}.$$

Two classes of solutions of this equation exist:

i) *W* is negative, the corresponding eigenfunctions are square integrable and the values of *W* are quantized (discrete spectrum).

ii) *W* is non-negative. Every real non-negative value of *W* is physically allowed (continuous spectrum), the corresponding eigenfunctions are non-square integrable.

The wavefunctions are known as bound states, in contrast to the class (ii) solutions that are known as *scattering states.*

For negative *W* the quantity $\alpha \equiv 2\sqrt{-2W}$ is real and positive. The scaling of *y*, i.e., substitution of $x \equiv \alpha y$ gives the Schrödinger equation:

$$\left[\frac{d^2}{dx^2} - \frac{l(l+1)}{x^2} + \frac{2}{\alpha x} - \frac{1}{4} \right] u_l = 0, \text{ with } x \geq 0.$$

For $x \to \infty$, the inverse powers of *x* are negligible and a solution for large *x* is $\exp(-x/2)$. The other solution, $\exp(x/2)$, is physically non-acceptable. For $x \to 0$, the inverse square power dominates and a solution for small *x* is x^{l+1}. The other solution, x^{-l}, is physically non-acceptable. Hence, to obtain a full range solution we substitute:

$$u_l(x) = x^{l+1} e^{-x/2} f_l(x).$$

The equation for $f_l(x)$ becomes,

$$\left[x\frac{d^2}{dx^2} + (2l+2-x)\frac{d}{dx} + (v-l-1) \right] f_l(x) = 0 \text{ with } v = (-2W)^{-\frac{1}{2}}.$$

Provided $v-l-1$ is a non-negative integer, say *k*, this equation has well-behaved (regular at the origin, vanishing for infinity) polynomial solutions written as:

$$L_k^{(2l+1)}(x), \qquad k = 0,1,\ldots,$$

which are generalized Laguerre polynomials of order k. We will take the convention for generalized Laguerre polynomials of Abramowitz and Stegun.

The energy becomes,

$$W = -\frac{1}{2n^2}, \text{ with } n \equiv k + l + 1.$$

The principal quantum number n satisfies $n \geq l + 1$, or $l \leq n - 1$. Since α = 2/n, the total radial wavefunction is,

$$R_{nl}(r) = N_{nl}\left(\frac{2Zr}{na_0}\right)^l e^{-\frac{Zr}{na_0}} L_{n-l-1}^{(2l+1)}\left(\frac{2Zr}{na_0}\right),$$

with normalization constant,

$$N_{nl} = \left[\left(\frac{2Z}{na_0}\right)^3 \cdot \frac{(n-l-1)!}{2n[(n+l)!]}\right]^{\frac{1}{2}},$$

and energy,

$$E_n = -\frac{Z^2}{2n^2}E_h, \qquad n = 1, 2, \ldots.$$

In the computation of the normalization constant use was made of the integral.

$$\int_0^\infty x^{2l+2} e^{-x} [L_{n-l-1(x)}^{(2l+1)}]^2 dx = \frac{2n(n+l)l}{(n-l-1)!}.$$

Caveat on Completeness of Hydrogen-like Orbitals

In quantum chemical calculations hydrogen-like atomic orbitals cannot serve as an expansion basis, because they are not complete. The non-square-integrable continuum (E > 0) states must be included to obtain a complete set, i.e., to span all of one-electron Hilbert space.

List of Radial Functions

The following list of radial functions $R_{nl}(r)$. The scaled distance is,

$$\rho_n \equiv \frac{2Zr}{a_0 n}.$$

$$R_{10}(r) = \left(\frac{Z}{a_0}\right)^{3/2} e^{-\rho_n/2} 2$$

$$R_{20}(r) = \left(\frac{Z}{a_0}\right)^{3/2} e^{-\rho_n/2} \frac{1}{2\sqrt{2}}(2 - \rho_n)$$

$$R_{21}(r)=\left(\frac{Z}{a_0}\right)^{3/2} e^{-\rho_n/2}\frac{1}{2\sqrt{6}}\rho_n$$

$$R_{30}(r)=\left(\frac{Z}{a_0}\right)^{3/2} e^{-\rho_n/2}\frac{1}{9\sqrt{3}}(6-6\rho_n+\rho_n^2)$$

$$R_{31}(r)=\left(\frac{Z}{a_0}\right)^{3/2} e^{-\rho_n/2}\frac{1}{9\sqrt{6}}(4-\rho_n)\rho_n$$

$$R_{32}(r)=\left(\frac{Z}{a_0}\right)^{3/2} e^{-\rho_n/2}\frac{1}{9\sqrt{30}}\rho_n^2$$

$$R_{40}(r)=\left(\frac{Z}{a_0}\right)^{3/2} e^{-\rho_n/2}\frac{1}{96}(24-36\rho_n+12\rho_n^2-\rho_n^3)$$

$$R_{41}(r)=\left(\frac{Z}{a_0}\right)^{3/2} e^{-\rho_n/2}\frac{1}{32\sqrt{15}}(20-10\rho_n+\rho_n^2)\rho_n$$

$$R_{42}(r)=\left(\frac{Z}{a_0}\right)^{3/2} e^{-\rho_n/2}\frac{1}{96\sqrt{5}}(6-\rho_n)\rho_n^2$$

$$R_{43}(r)=\left(\frac{Z}{a_0}\right)^{3/2} e^{-\rho_n/2}\frac{1}{96\sqrt{35}}\rho_n^3$$

$$R_{50}(r)=\left(\frac{Z}{a_0}\right)^{3/2} e^{-\rho_n/2}\frac{1}{300\sqrt{5}}(120-240\rho_n+120\rho_n^2-20\rho_n^3+\rho_n^4)$$

$$R_{51}(r)=\left(\frac{Z}{a_0}\right)^{3/2} e^{-\rho_n/2}\frac{1}{150\sqrt{30}}(120-90\rho_n+18\rho_n^2-\rho_n^3)\rho_n$$

$$R_{52}(r)=\left(\frac{Z}{a_0}\right)^{3/2} e^{-\rho_n/2}\frac{1}{150\sqrt{70}}(42-14\rho_n+\rho_n^2)\rho_n^2$$

$$R_{53}(r)=\left(\frac{Z}{a_0}\right)^{3/2} e^{-\rho_n/2}\frac{1}{300\sqrt{70}}(8-\rho_n)\rho_n^3$$

$$R_{54}(r)=\left(\frac{Z}{a_0}\right)^{3/2} e^{-\rho_n/2}\frac{1}{900\sqrt{70}}\rho_n^4$$

References

- Hydrogen: newworldencyclopedia.org, Retrieved 20 March, 2019
- Deuterium: newworldencyclopedia.org, Retrieved 30 May, 2019
- Tritium: newworldencyclopedia.org, Retrieved 14 August, 2019
- Hydrogen-like-atom: theochem.ru.nl, Retrieved 21 June, 2019

4

Multi-electron Atoms

The atoms which have more than one electron are called multi-electron atoms. The presence of more than one electron in the atom leads to phenomena like shielding. This chapter closely examines the key concepts of multi-electron atoms such as helium to provide an extensive understanding of the subject.

Atom with more than one electrons, such as Helium (He), and Nitrogen (N), are referred to as multi-electron atoms. Hydrogen is the only atom in the periodic table that has one electron in the orbitals underground state.

First, electrons repel against each other. Particles with the same charge repel each other, while oppositely charged particles attract each other. For example, a proton, which is positively charged, is attracted to electrons, which are negatively charged. However, if we put two electrons together or two protons together, they will repel one another. Since neutrons lack a charge, they will neither repel nor attract protons or electrons.

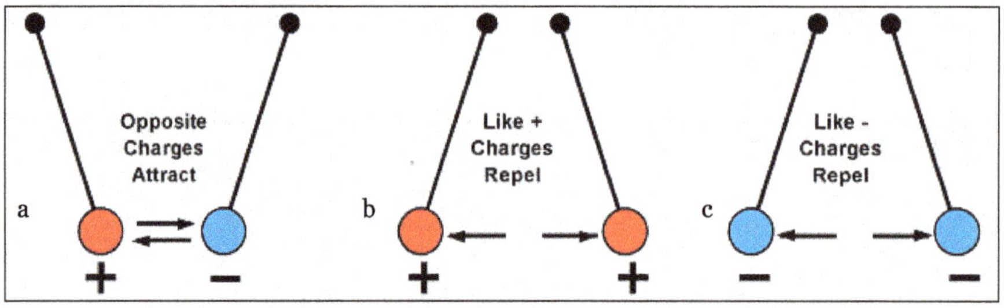

a) When oppositely charged particles, an electron and a proton, are placed together, they attract each other. b) The two protons are repelling each other for the same reason. c) The two electrons are placed together and repel each other because of the same charge.

Protons and neutrons are located in an atom's nucleus. Electrons float around the atom in energy levels. Energy levels consist of orbitals and sub-orbitals. The lower the energy level the electron is located at, the closer it is to nucleus. As we go down and to the right of the periodic table, the number of electrons that an element has increases. Since there are more electrons, the atom experiences greater repulsion and electrons will tend to stay as far away from each other as possible. Our main focus is what effects take place when more electrons surround the nucleus.

Shielding (Screening)

With more protons in the nucleus, the attractive force for electrons to the nucleus is stronger. Thus, the orbital energy becomes more negative (less energy). Orbital energy also depends on the type of l orbital an electron is located in. The lower the number of l, the closer it is to the nucleus. For example, l = 0 is the s orbital. S orbitals are closer to the nucleus than the p orbitals (l = 1) that are closer to the nucleus than the d orbitals (l = 2) that are closer to the f orbitals (l = 3).

More electrons create the shielding or screening effect. Shielding or screening means the electrons closer to the nucleus block the outer valence electrons from getting close to the nucleus. Imagine being in a crowded auditorium in a concert. The enthusiastic fans are going to surround the auditorium, trying to get as close to the celebrity on the stage as possible. They are going to prevent people in the back from seeing the celebrity or even the stage. This is the shielding or screening effect. The stage is the nucleus and the celebrity is the protons. The fans are the electrons. Electrons closest to the nucleus will try to be as close to the nucleus as possible. The outer/valence electrons that are farther away from the nucleus will be shielded by the inner electrons. Therefore, the inner electrons prevent the outer electrons from forming a strong attraction to the nucleus. The degree to which the electrons are being screened by inner electrons can be shown by ns < np < nd < nf where n is the energy level. The inner electrons will be attracted to the nucleus much more than the outer electrons. Thus, the attractive forces of the valence electrons to the nucleus are reduced due to the shielding effects. That is why it is easier to remove valence electrons than the inner electrons. It also reduces the nuclear charge of an atom.

Penetration

Penetration is the ability of an electron to get close to the nucleus. The penetration of ns > np > nd > nf. Thus, the closer the electron is to the nucleus, the higher the penetration. Electrons with higher penetration will shield outer electrons from the nucleus more effectively. The s orbital is closer to the nucleus than the p orbital. Thus, electrons in the s orbital have a higher penetration than electrons in the p orbital. That is why the s orbital electrons shield the electrons from the p orbitals. Electrons with higher penetration are closer to the nucleus than electrons with lower penetration. Electrons with lower penetration are being shielded from the nucleus more.

Radial Probability Distribution

Radial probability distribution is a type of probability to find where an electron is mostly likely going to be in an atom. The higher the penetration, the higher probability of finding an electron near the nucleus. As shown by the graphs, electrons of the s orbital are found closer to the nucleus than the p orbital electrons. Likewise, the lower the energy level an electron is located at, the higher chance it has of being found near the nucleus. The smaller the energy level (n) and the orbital angular momentum quantum number (l) of an electron is, the more likely it will be near the nucleus. As electrons get to higher and higher energy levels, the harder it is to locate it because the radius of the sphere is greater. Thus, the probability of locating an electron will be more difficult. Radial probability distribution can be found by multiplying $4\pi r^2$, the area of a sphere with a radius of r and $R^2(r)$.

Radial Probability Distribution = $4\pi r^2$ X $R^2(r)$

By using the radial probability distribution equation, we can get a better understanding about an electron's behavior, as shown in figures.

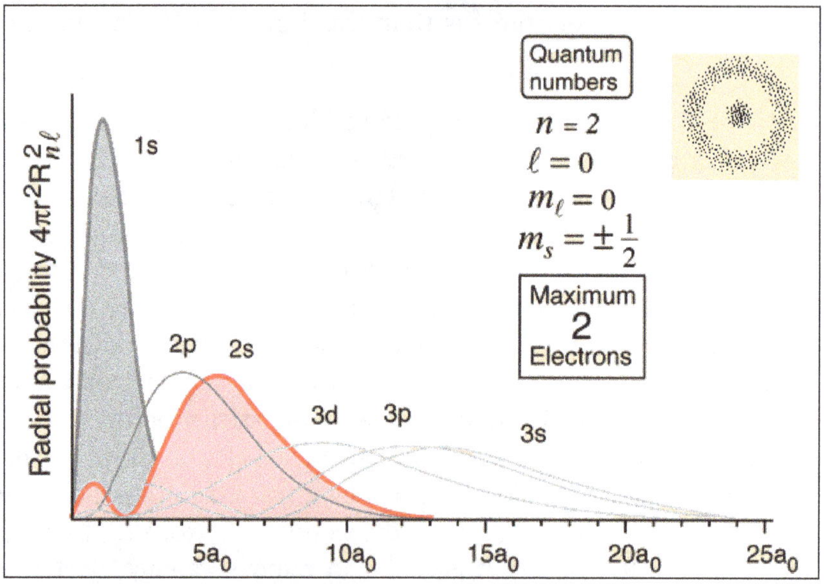

Figures show that the lower the energy level, the higher the probability of finding the electron close to the nucleus. Also, the lower momentum quantum number gets, the closer it is to the nucleus.

Effective Nuclear Charge (Z_{eff})

Effective nuclear charge (Z_{eff}) is the positive nuclear charge that is experienced by an electron. The more electrons shielding an electron has the lower the Z_{eff} value. Since it is farther away from the nucleus, the force is weaker. Using the analogy of the stage, imagine you are in the back. With all the noise and shielding from the inner people, you can barely see the stage or even hear the celebrity. Likewise, the outer electrons will have a lower value of Z_{eff}. Thus, the lower the Z_{eff} value, the less attracted the electron is to the nucleus. So, it is not held tightly by the nucleus. As a result, the electron has to be in a higher energy level orbital because it farther away from the nucleus. Therefore, electrons in the lower energy s orbital with higher penetration are less shielded by other electrons and experience a higher Z_{eff} than p orbital electrons. That is why orbitals have different energy. 1s electrons have less energy than 2s electrons. 1s electrons shield the 2s electrons from "seeing" the nucleus because it has a higher penetration.

$$Z_{eff} = Z - S$$

where,

- Z_{eff} is effective nuclear charge.
- Z is the atomic number.
- S is the number of inner energy level/core electrons.

For example: Z_{eff} of Li = 3 - 2 = 1.

Effect of Effective Nuclear Charge and Energy Level on the Orbital Energy

$$En \propto \frac{-Z_{eff}^2}{n^2}$$

where,

- En is the orbital energy.
- Z_{eff} is the effective nuclear charge.
- n is the energy level.

Z_{eff} is directly proportional to En and is inversely proportional to n. As Z_{eff} increases, En increases. As n increases, En decreases.

Almost all atoms are multiple-electron atoms and their description is more complicated due the increase in the number of different interactions. Consider an atom with atomic number Z. This atom will contain a nucleus with a charge Ze and Z electrons. To describe this system we need to consider the interactions between each electron and the nucleus and the interactions between the electrons. In order to describe such atoms, we will start with dealing with the strongest interactions first. The description is complicated by the fact that the electrons are identical particles. In Classical Physics, identical particles can be distinguished on the basis of careful measurements. However, in Quantum Physics, electrons cannot be distinguished unless we disturb them. First consider an atom with two electrons with no mutual interactions. The Schrödinger equation describing this system is:

$$-\frac{\hbar^2}{2m}\nabla_1^2\psi_T - \frac{\hbar^2}{2m}\nabla_2^2\psi_T + V_T\psi_T = E_T\psi_T$$

The position of particle 1 is (x_1, y_1, z_1); the position of particle 2 is (x_2, y_2, z_2). If we assume there is no interaction between the electrons we can write the interaction potential as,

$$V_T(x_1, y_1, z_1, x_2, y_2, z_2) = V_1(x_1, y_1, z_1) + V_2(x_2, y_2, z_2)$$

and the Schrödinger equation becomes,

$$-\frac{\hbar^2}{2m}\nabla_1^2\psi_T - \frac{\hbar^2}{2m}\nabla_2^2\psi_T + V_1\psi_T + V_2\psi_T = E_T\psi_T$$

Consider the possibility that the wavefunction is the product of two wavefunctions: one associated with the particle 1 and one associated with particle 2:

$$\psi_T(1,2) = \psi_\alpha(1)\psi_\beta(2)$$

To determine if we can rewrite the wavefunction in this form, we need to substitute it in the Schrödinger equation:

$$\left(-\frac{\hbar^2}{2m}\nabla_1^2\psi_\alpha\right)\psi_\beta + \left(-\frac{\hbar^2}{2m}\nabla_2^2\psi_\beta\right)\psi_\alpha + V_1\psi_\alpha\psi_\beta + V_2\psi_\alpha\psi_\beta = E_T\psi_\alpha\psi_\beta$$

Dividing this equation by the wavefunction we obtain,

$$\underbrace{\left(-\frac{\hbar^2}{2m}\frac{1}{\psi_\alpha}\nabla_1^2\psi_\alpha+V_1\right)}_{E_1}+\underbrace{\left(-\frac{\hbar^2}{2m}\frac{1}{\psi_\beta}\nabla_2^2\psi_\beta+V_2\right)}_{E_2}\psi_\alpha=E_T$$

In order for this equation to be satisfied for all positions of particles 1 and 2, the terms between the parentheses need to be constant. This requires that,

$$-\frac{\hbar^2}{2m}\frac{1}{\psi_\alpha}\nabla_1^2\psi_\alpha+V_1\psi_\alpha=E_1\psi_\alpha$$

and

$$-\frac{\hbar^2}{2m}\frac{1}{\psi_\alpha}\nabla_2^2\psi_\beta+V_2\psi_\beta=E_2\psi_\beta$$

Since the V is the Coulomb potential, the wavefunctions ψ_α and ψ_β are one-electron wavefunctions. The probability density distribution is,

$$\psi^*_T\psi_T=\left(\psi_\alpha(1)\psi_\beta(2)\right)^*\left(\psi_\alpha(1)\psi_\beta(2)\right)=\left(\psi^*_\alpha(1)\psi_\alpha(1)\psi^*_\beta(2)\psi_\beta(2)\right)$$

In general, we cannot assume that the energies of the two states are the same, and the value of the wave function $\psi_\alpha(1)$ will be different from the value of the wavefunction $\psi_\beta(1)$. The probability density distribution will thus change when we exchange particles 1 and 2:

$$\left(\psi^*_\alpha(1)\psi_\alpha(1)\right)\left(\psi^*_\beta(2)\psi_\beta(2)\right)\neq\left(\psi^*_\alpha(2)\psi_\alpha(2)\right)\left(\psi^*_\beta(1)\psi_\beta(1)\right)$$

Since the probability density distribution changes when we exchange particles 1 and 2, we say that the particles are not indistinguishable. We thus say that the way we combined the two wave functions is not proper and the proposed total wave function is not a proper eigenfunction of the Schrödinger equation for a two-electron atom. There are however different ways of combining the two wave functions. For example, consider the following total wave functions:

$$\psi_s(1,2)=\frac{1}{2}\sqrt{2}\left\{\psi_\alpha(1)\psi_\beta(2)+\psi_\alpha(1)\psi_\alpha(2)\right\}$$

$$\psi_A(1,2)=\frac{1}{2}\sqrt{2}\left\{\psi_\alpha(1)\psi_\beta(2)-\psi_\beta(1)\psi_\alpha(2)\right\}$$

The numerical factor in these expressions ensures that the total wave functions are normalized:

$$\begin{aligned}\left\langle\psi_s(1,2)\middle|\psi_s(1,2)\right\rangle&=\frac{1}{2}\left\langle\psi_\alpha(1)\psi_\beta(2)+\psi_\beta(1)\psi_\alpha(2)\middle|\psi_\alpha(1)\psi_\beta(2)+\psi_\alpha(2)\right\rangle\\&=\frac{1}{2}\left\langle\psi_\alpha(1)\psi_\beta(2)\middle|\psi_\alpha(1)\psi_\beta(2)\right\rangle+\frac{1}{2}\left\langle\psi_\alpha(1)\psi_\beta(2)\middle|\psi_\beta(1)\psi_\alpha(2)\right\rangle\end{aligned}$$

$$+\frac{1}{2}\langle\psi_\beta(1)\psi_\alpha(2)|\psi_\alpha(1)\psi_\beta(2)\rangle+\frac{1}{2}\langle\psi_\beta(1)\psi_\alpha(2)|\psi_\beta(1)\psi_\alpha(2)\rangle$$
$$=\frac{1}{2}\langle\psi_\alpha(1)\psi_\beta(2)|\psi_\alpha(1)\psi_\beta(2)\rangle+\frac{1}{2}\langle\psi_\beta(1)\psi_\alpha(2)|\psi_\beta(1)\psi_\alpha(2)\rangle$$
$$=\frac{1}{2}+\frac{1}{2}=1$$

These wavefunctions are symmetric and asymmetric under the exchange of particles 1 and 2:

$$\psi_s(2,1)=\frac{1}{2}\sqrt{2}\{\psi_\alpha(2)\psi_\beta(1)+\psi_\beta(2)\psi_\alpha(1)\}=\psi_s(1,2)$$
$$\psi_A(2,1)=\frac{1}{2}\sqrt{2}\{\psi_\alpha(2)\psi_\beta(1)-\psi_\beta(2)\psi_\alpha(1)\}=-\psi_A(1,2)$$

Their probability density distributions do not change under particle exchange:

$$\psi_s^*(2,1)\psi_s(2,1)=\psi_s^*(1,2)\psi_s(1,2)$$
$$\psi_A^*(2,1)\psi_A(2,1)=[-\psi_A(1,2)]^*[-\psi_A(1,2)]=\psi_A^*(1,2)\psi_A(1,2)$$

We note that the energy of both wave functions is the same,

$$E_s=E_\alpha+E_\beta=E_A$$

and the symmetric and anti-symmetric states are thus degenerate states. The average distance between the two particles will be different for the two solutions. Now consider what will happen when the two electrons have the same quantum numbers: $\alpha=\beta$. In this case, the wave functions will be equal to:

$$\psi_s(1,2)=\frac{1}{2}\sqrt{2}\{\psi_\alpha(1)\psi_\alpha(2)+\psi_\alpha(1)\psi_\alpha(2)\}=\sqrt{2}\psi_\alpha(1)\psi_\alpha(2)$$
$$\psi_A(1,2)=\frac{1}{2}\sqrt{2}\{\psi_\alpha(1)\psi_\alpha(2)-\psi_\alpha(1)\psi_\alpha(2)\}=0$$

We thus say that if the wavefunction describing the system is asymmetric, the particles cannot have the same quantum numbers. When the wavefunction describing the system is symmetric, the particles can have the same quantum numbers.

In 1925, Pauli concluded on the basis of his study of atomic structure that in multi-electron atoms no more than one electron can be in a given quantum state. This is called the Pauli exclusion principle. An alternative expression of the exclusion principle is the statement that the wavefunction describing a multiple-electron system must be asymmetric.

To construct asymmetric wavefunctions for multiple-electron atoms we can use the Slater Determinant:

$$\psi_A=\frac{1}{\sqrt{N!}}\begin{vmatrix}\psi_\alpha(1) & \psi_\alpha(2) & \cdot\,\cdot \\ \psi_\beta(1) & \psi_\beta(2) & \\ \cdot & & \\ \cdot & & \end{vmatrix}$$

Adding spin to the system increases the number of possible wavefunctions. To determine the number of possible ways of combining spin, let us start with a two-electron system:

- The spin of each electron is ½. Each electron can be in a state with m_s = +½ and a state with m_s = -½.
- The spin wavefunction can be represented by a symmetric and an asymmetric wavefunction:

$$\psi_{A,spin} = \frac{1}{2}\sqrt{2}\{(+1/2,-1/2)-(-1/2,+1/2)\}\, m_s = 0$$

$$\psi_{s,spin} = \begin{Bmatrix} (+1/2,+1/2) & m_s = 1 \\ \frac{1}{2}\sqrt{2}\{(+1/2,-1/2)+(-1/2,+1/2)\} & m_s = 0 \\ (-1/2,-1/2) & m_s = -1 \end{Bmatrix}$$

- The symmetric and asymmetric wavefunctions corresponds to two values of the total spin of the two spin-½ particle system:
    - S_{12} = 1: Triplet State, Symmetric, $\psi_{S,spin}$.
    - S_{12} = 0: Singlet State, Asymmetric, $\psi_{A,spin}$.

Since the total wave function of the multi-electron atom must be asymmetric, the spin wave function will constrain the spatial wave function:

$$\psi_{multi-electron} = \psi_{spatial}\,\psi_{spin}$$

Consider the following two possible configurations for a two-electron atom:

- Singlet spin state: When the system is in the singlet spin state, its spin wave function will be asymmetric. The spatial wave function must thus be symmetric.

$$\psi_{spatial}(1,2) = \frac{1}{2}\sqrt{2}\left\{\psi_\alpha(1)\psi_\beta(2)+\psi_\beta(1)\psi_\alpha(2)\right\}$$

  When the two electrons are in the same spatial state, α = β, the probability density distribution will not be equal to zero even if the electrons are located at the same position (1 = 2). For a system in the singlet spin state, we can find the two electrons close together.

- Triplet spin state: When the system is in the triplet spin state, its spin wave function is symmetric. The spatial wave function must thus be asymmetric.

$$\psi_{spatial}(1,2) = \frac{1}{2}\sqrt{2}\left\{\psi_\alpha(1)\psi_\beta(2)+\psi_\beta(1)\psi_\alpha(2)\right\}$$

  If electrons 1 and 2 are located at the same position the spatial wave function is zero and the probability density distribution is thus also equal to 0. For a system in the triplet spin state, the two electrons cannot be located at the same position. In the triplet state it appears as if the electrons repel each other.

The preceding discussion shows that there is a coupling between the spin and the space variables. The electrons act as if a force is acting between them and the force depends on the relative orientation of their spin. This force is called the exchange force. The effect is shown schematically in the figure below. The effect of the exchange force on the energy levels of Helium is shown in the figure at the bottom of the page. The figure shows the following features:

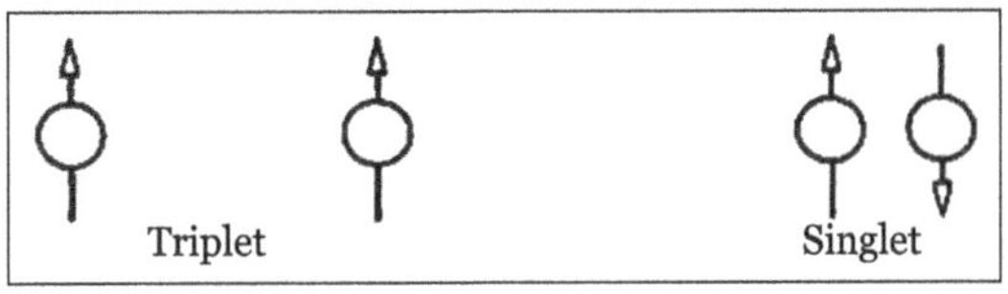

- The energy levels on the left are the energy levels that would be found if there is no Coulomb interaction between the electrons.
- The energy levels in the center are the energy levels that are found when the Coulomb interaction between the electrons is included. We see that the n = 1, n = 2 level is split. This is a result of the fact that the distance between the two electrons is somewhat larger when one electron is in the $(n=1, \ell=0)$ state and the other one is in the $(n=2, \ell=0)$ state compared to the situation in which one electron is in the $(n=1, \ell=0)$ state and the other one is in the $(n=2, \ell=1)$ state. The Coulomb repulsion between the two electrons will be less in former compared to the latter configuration. The level with $\ell=1$ is thus located at a slightly higher energy that the level with $\ell=0$.

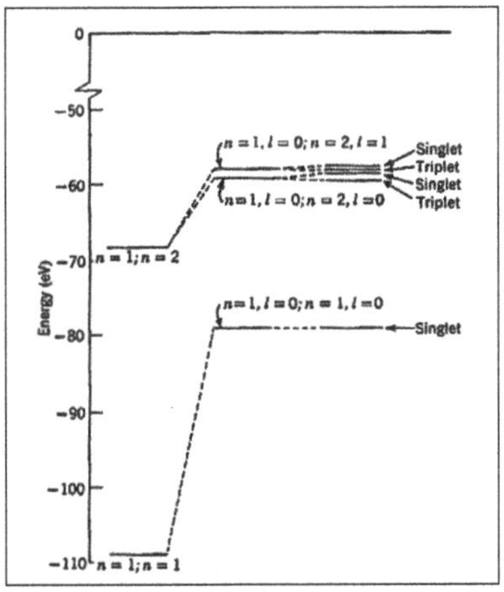

- The levels on the right-hand side include the exchange force. Since the average distance between the electrons in the triplet state is larger than the average distance between the electrons in the singlet state, the energy of the triplet state will be less than the energy of the singlet state.

In order to describe multi-electron atoms we use the Hartree theory. This theory relies on multiple approximations and including gradually less important interactions. In order to describe a multi-electron atom we will start with considering the following forces:

- The attractive forces between the electrons and the nucleus. This force increases when r decreases.

- The repulsive forces between the electrons. This force decreases when r decreases. The reason that the force decreases is due to the fact that when the electron is located at small r the repulsive forces due to all other electrons add up to zero. When the electron is located near the edge of the atom, the repulsive forces do not cancel, but provide a net force point away from the center of the atom.

To describe the motion of the electrons we make the following assumptions:

- The atom has Z electrons and a nucleus with a positive charge Ze.
- The electrons move independently.
- The effect of the presence of other electrons is included as a modification to the potential seen by the electron.
- Since the atom has spherical symmetry it is assumed that the potential also has spherical symmetry.
- When r approaches 0, it is assumed that the potential approaches.

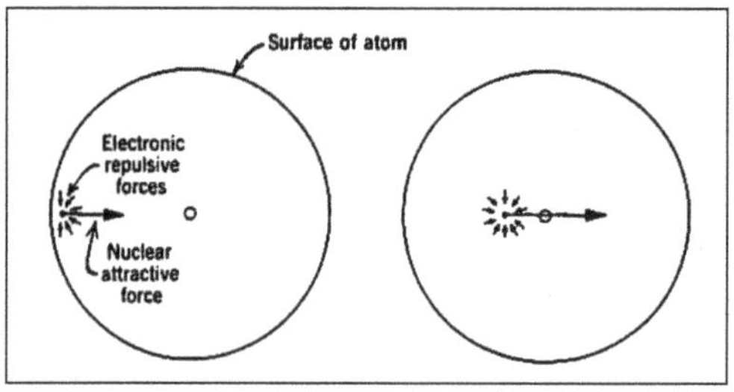

$$V \underset{r\to 0}{\to} -\frac{1}{4\pi\varepsilon_0}\frac{Z_e^2}{r}$$

The net charge seen by the electron is Ze.

- When r approaches infinity, it is assumed that the potential approaches

$$V \underset{r\to\infty}{\to} -\frac{1}{4\pi\varepsilon_0}\frac{e^2}{r}$$

The net charge seen by the electron is e which is the sum of the nuclear charge Ze and the charge due to Z-1 electrons.

The following procedure is followed in order to determine the wavefunctions that describes the multi-electron atom:

- Construct the potential V(r) using the procedure outlined above.
- Solve the time-independent Schrödinger equation and determine the eigenfunctions.
- Construct the ground state of the atom by filling the eigenfunctions obtained in step 2. If there are Z electrons in the atom, we will use the first Z eigenfunctions.

- Calculate the spatial charge distribution by calculating the following charge density distribution for each electron: $-e\langle\psi|\psi\rangle$.
- Using the spatial charge distribution obtained in step 4, calculate a modified potential and go back to step 2.
- Stop the cycle if the potential V converges. This also implies that ψ converges.

The exclusion principle is used in step 3 but the asymmetric wavefunction requirement is not used.

When we follow these steps we can determine the wavefunction of the ground state of the multielectron atom. Based on this wavefunction we can come to the following conclusions:

- The wave function of the ground state can be written in the following way:

$$\Psi_{n\ell m_\ell m_s} = R_{n\ell}(r) \underbrace{\Theta_{\ell m_\ell}(\theta)\Phi_{m\ell}(\varphi)}_{\text{Same functions as were used for single-electron atoms}} \underbrace{(m_s)}_{\text{Spin eigenfunction}}$$

- The sum of the solutions for a given combination of n and ℓ is spherically symmetric. For example, the probability density distribution for $n = 2$ and $\ell = 1$ is equal to,

$$\psi^*_{2,1}\psi_{2,1} = \frac{1}{3}\left(\psi^*_{2,1,-1}\psi_{2,1,-1} + \psi^*_{2,1,0}\psi_{2,1,0} + \psi^*_{2,1,1}\psi_{2,1,1}\right)$$

$$= \frac{1}{96\pi}\left(\frac{Z}{a_0}\right)^3 e^{-zr/a_0}\left(\frac{Zr}{a_0}\right)^2$$

  The probability density distribution does not depend on the polar and azimuthal angle. It thus has spherical symmetry. The asymmetry of the charge distribution is due to the electrons in the highest energy state that is not fully occupied.

- The r dependence of the wave function will be different for multi-electron atoms. For a given n, electrons are found in a limited region of space.
- The spherical charge distributions imply that the modified potentials are also spherically symmetric:

$$V_n(r) = -\frac{1}{4\pi\varepsilon_0}\frac{Z_n e^2}{r}$$

  In this equation, n is the principle quantum number and Z_n is the effective charge. It is found that Z_n has the following values:

    - For the innermost shell $(n=1)$: $Z_n = Z - 2$.
    - For the next shell $(n=2)$: $Z_n = Z - 10$. Note: $10 = 2 + 6 + 2$.
    - For the outermost shell: $Z_n = n$.

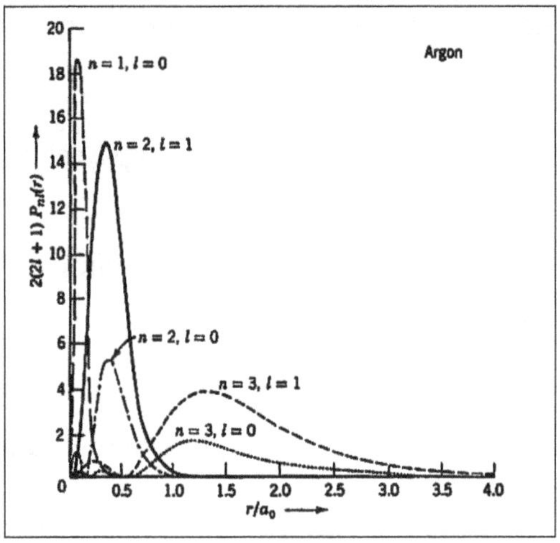

Argon

- The radius of the orbits of the n = 1 electrons decreases with increasing Z:

$$\bar{r} \simeq \frac{a_0}{Z} \Rightarrow \bar{r}_1 \simeq \frac{a_0}{Z_1} \simeq \frac{a_0}{Z-2} = \frac{\bar{r}_{1,H}}{Z-2}$$

- The energy of the innermost electrons becomes more negative with increasing Z:

$$E_1 = \frac{\mu Z_1^2 e^4}{(4\pi\varepsilon_0)^2(2\hbar^2 n^2)} = Z_1^2\left\{-\frac{\mu e^4}{(4\pi\varepsilon_0)^2(2\hbar^2)}\right\} = (Z-2)^2 E_{1,H}$$

The binding energy of the innermost electron in the Hydrogen atom is -13.6 eV. If Z is large enough, the binding energy will exceed two times the rest mass of the electron. This happens when,

$$(Z-2)^2 = \frac{E_1}{E_{1,H}} = \frac{2\times 511{,}000}{13.6} = 75{,}147 \Rightarrow Z = 276$$

- The energy of the electrons in the outermost shell is equal to,

$$E_n = -\frac{\mu Z_n^2 e^4}{(4\pi\varepsilon_0)(2\hbar^2 n^2)} = -\frac{\mu n^2 e^4}{(4\pi\varepsilon_0)^2(2\hbar^2 n^2)} = -\frac{\mu e^4}{(4\pi\varepsilon_0)^2(2\hbar^2)} = E_{1,H}$$

which is the energy of the ground state of the Hydrogen atom.

- Electrons in the outermost shell move in an orbit with an average radius r equal to,

$$\bar{r} = \frac{n^2 a_0}{Z} \simeq \frac{n^2 a_0}{n} = n a_0$$

Since n increases slowly when Z increases (for H n = 1, for Ne n – 2, etc.) the radius of the atom will only slowly increase with increasing Z. For example, n doubles when Z increases by a factor of 10.

- For small *r*, the probability density distribution is proportional to $r^{2\ell}$. The distribution in this region is not sensitive to the details of the potential. Since $r < 1$, the probability of finding a state with a high angular momentum at a given r is smaller than the probability of finding a state with a low angular momentum at that r. Within a given n shell, the most negative state is the $\ell = 0$ state; the next state is the $\ell = 1$ state, etc.

The orbits of electrons in multi-electron atoms are specified by using the so-called spectroscopic notation $n\ell$. The key ingredients of the spectroscopic notation are the principle quantum number n and angular momentum quantum number ℓ. States with different angular momentum quantum numbers are specified in the following way:

ℓ	Spectroscopic Notation
0	s
1	p
2	d
3	f
4	g
5	h
6	i

In terms of increasing energy, the sub-shells are arranged in the following way:

$$1s, 2s, 2p, 3s, 3p, 4s, 3d, ...$$

The rules for ordering the sub-shells are as follows:

- For a given outer sub-shell, lowest ℓ, lowest E.
- For a given ℓ, the outer sub-shell with the lowest n has the lowest E.

The relative energies of sub-shells depend on the atomic number of the atom. For certain values of Z the ordering of the sub-shells will change.

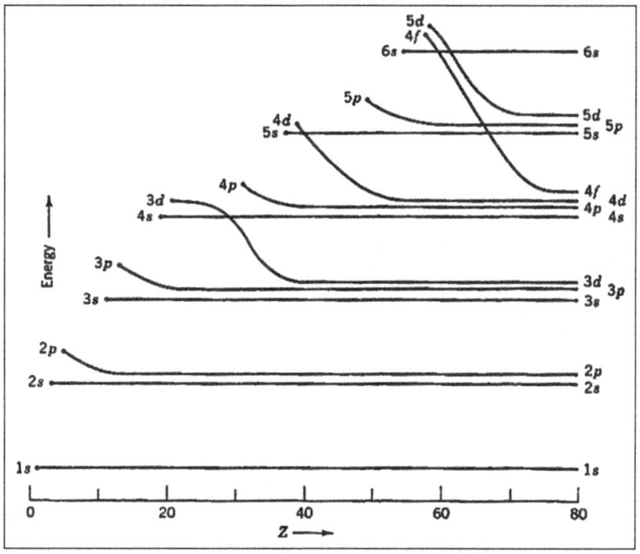

Based on the energy of the levels shown in the figure, additional rules for subshells are observed:

- Every p shell has a higher energy than the preceding s and d shells.
- Every s shell has a higher energy than the preceding p shell.

Based on our current understanding of the shell and sub-shell structure of multi-electron atoms we can:

- Understand the properties of noble gasses. Noble gasses have filled outer p shell. Because of the large gap between the p shell and the next s shell, the first excited state is far above the ground state and the atom is difficult to excite. The effect is a high ionization energy for noble gasses, as shown in the figure on the top of this page. As a result of the shell structure, the noble gas atoms are spherical. The charges of the noble atoms do not produce an electric field outside the atom, and noble gas atoms do not interact chemically with other atoms.
- Understand the ionization energies, electron affinity, etc. These properties are related the properties of the last electron.
- A group of atoms, called the alkalis, have one loosely bound electron in the outer s shell. This loosely-bound electron is easy to remove and as a result, these atoms have a low ionization energy.
- Another group of atoms, called the halogens, have one less electron than is required to fill their outer p shell. As a result, these atoms are very likely to pick up an electron to fill their outer p shell. The halogens thus have a high electron affinity.

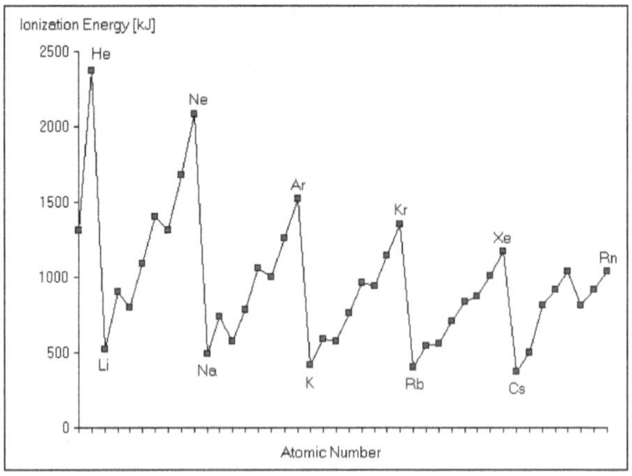

The chemical properties of atoms are primarily determined by the properties of the electrons in the outer shells. In order to determine the properties of electrons in the inner shells, X rays can be used. When energetic electrons interact with atoms, they can eject electrons from one of the inner shells and create a hole in this shell. This leaves the atom in a highly excited state. Consider the following scenario:

- An electron from the 1s shell is removed.
- The hole created in the 1s shell is filled with an electron from the 2p shell. When this happens, a photon is emitted with an energy equal to the energy difference of the 2p shell and the 1s shell.

- The hole in the 2p shell is filled with an electron from the 3d shell. When this happens, a photon is emitted with an energy equal to the energy difference of the 3d shell and the 2p shell.
- The hole continues to move up to higher-energy states.

The removal of an electron from an inner shell thus produces a cascade of low energy photons, X-rays, which carry information about the energy structure of the inner shells of the atom. An example of an X-ray spectrum of Tungsten is shown in the figure below.

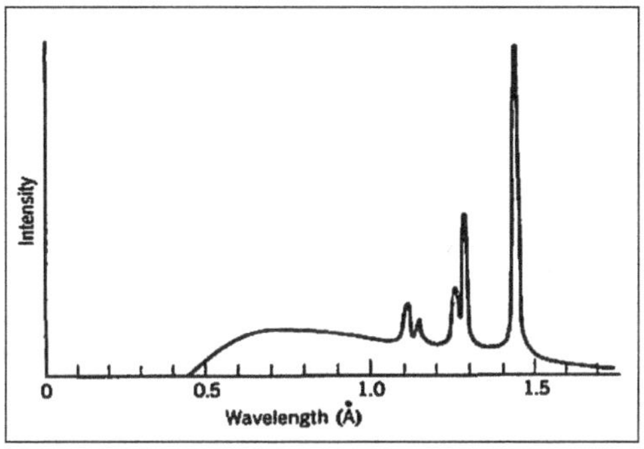

When we look at the transitions in the atom, it appears as if the hole is jumping from low lying states to high lying states. The transitions made by the hole satisfy selection rules similar to those satisfied by electron transitions in the outer shells:

$$\Delta \ell = \pm 1$$

$$\Delta j = 0, \pm 1$$

The energies of the X-rays changes smoothly as function of Z. For example the Ka lines have the following dependence on Z:

$$\frac{1}{\lambda} \simeq C(Z-a)^2$$

where a is a constant with a value between 1 and 2. A plot of the square-root of the inverse of the wavelength as function of Z should thus the a straight line. The slope of the line if the square-root of C and the line intercepts the horizontal axis at Z = a. An example of this type of data is shown in the figure on the right.

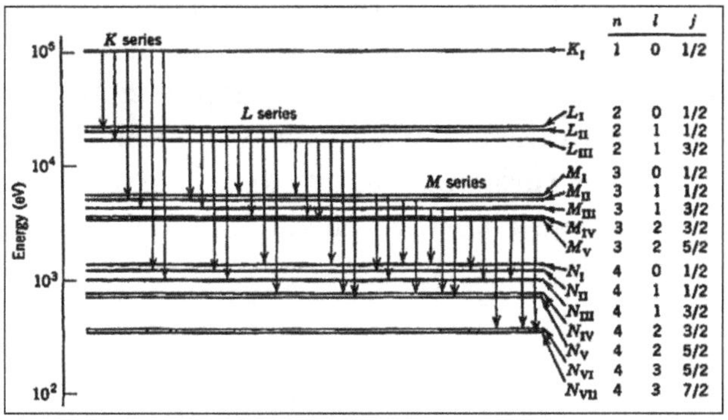

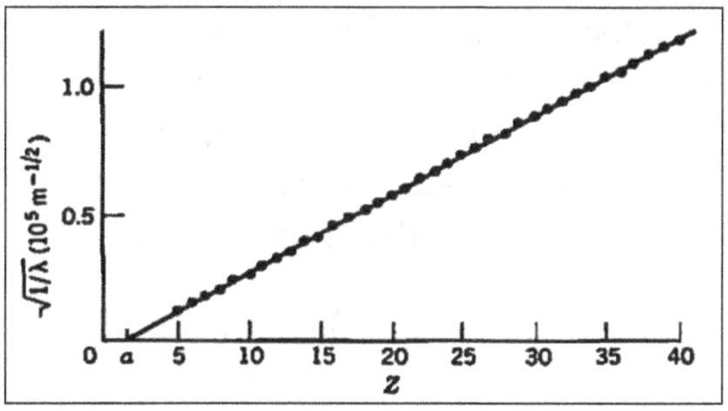

Helium Atom

The simplest multi electron system is the Helium atom. The second element in the periodic table provides our first example of a quantum-mechanical problem which cannot be solved exactly. Nevertheless, approximation methods applied to helium can give accurate solutions in perfect agreement with experimental results. In this sense, it can be concluded that quantum mechanics is correct for atoms more complicated than hydrogen. By contrast, the Bohr theory failed miserably in attempts to apply it beyond the hydrogen atom.

The helium atom has two electrons bound to a nucleus with charge Z = 2. The successive removal of the two electrons can be diagrammed as,

$$He \xrightarrow{I_1} He^+ + e^- \xrightarrow{I_2} He^{++} + 2e^-$$

The first ionization energy I_1, the minimum energy required to remove the first electron from helium, is experimentally 24.59 eV. The second ionization energy, I_2 is 54.42 eV. The last result can be calculated exactly since He^+ is a hydrogen-like ion. We have,

$$I_2 = -E_{1s}(\mathrm{He}^+) = \frac{Z^2}{2n^2} = 2\,\mathrm{hartrees} = 54.42\,\mathrm{eV}$$

The energy of the three separated particles on the right side of equation $He \xrightarrow{I_1} He^+ + e^- \xrightarrow{I_2} He^{++} + 2e^-$ is zero. Therefore the ground-state energy of helium atom is given by $E_0 = -(I_1 + I_2) = -79.02$ eV $= -2.90372$ hartrees. We will attempt to reproduce this value, as close as possible, by theoretical analysis.

Schrödinger Equation and Variational Calculations

The Schrödinger equation for He atom, again using atomic units and assuming infinite nuclear mass, can be written.

$$\left\{-\frac{1}{2}\nabla_1^2 - \frac{1}{2}\nabla_2^2 - \frac{Z}{r^2} + \frac{1}{r_{12}}\right\}\psi(r_1, r_2) = E\psi(r_1, r_2)$$

The five terms in the Hamiltonian represent, respectively, the kinetic energies of electrons 1 and 2, the nuclear attractions of electrons 1 and 2, and the repulsive interaction between the two electrons. It is this last contribution which prevents an exact solution of the Schrödinger equation and which accounts for much of the complication in the theory. In seeking an approximation to the ground state, we might first work out the solution in the absence of the $1/r_{12}$-term. In the Schrödinger equation thus simplified, we can separate the variables r_1 and r_2 to reduce the equation to two independent hydrogen-like problems. The ground state wavefunction (not normalized) for this hypothetical helium atom would be,

$$\psi(r_1, r_2) = \psi_{1s}(r_1)\psi_{1s}(r_2) = e^{-Z(r_1 + r_2)}$$

and the energy would equal 2 × ($-Z^2/2$) = − 4 hartrees, compared to the experimental value of −2.90 hartrees. Neglect of electron repulsion evidently introduces a very large error.

A significantly improved result can be obtained with the functional form, but with Z replaced by a adjustable parameter α, thus,

$$\bar{\psi}(r_1, r_2) = e^{-a(r_1 + r_2)}$$

Using this function in the variational principle, we have,

$$\bar{E} = \frac{\int \psi(r_1, r_2) \hat{H} \psi(r_1, r_2) d\tau_1 \tau_2}{\int \psi(r_1, r_2) \psi(r_1, r_2) d\tau_1 \tau_2}$$

where $\hat{H}$ is the full Hamiltonian as in equation $\left\{-\frac{1}{2}\nabla_1^2 - \frac{1}{2}\nabla_2^2 - \frac{Z}{r^2} + \frac{1}{r_{12}}\right\}\psi(r_1, r_2) = E\psi(r_1, r_2)$, including the $1/r_{12}$-term. The expectation values of the five parts of the Hamiltonian work out to,

$$\left\langle -\frac{1}{2}\nabla_1^2 \right\rangle = \left\langle -\frac{1}{2}\nabla_2^2 \right\rangle = \frac{\alpha^2}{2}$$

$$\left\langle -\frac{Z}{r_1} \right\rangle = \left\langle -\frac{Z}{r_2} \right\rangle = -Z\alpha, \left\langle \frac{1}{r_{12}} \right\rangle = \frac{5}{8}\alpha$$

The sum of the integrals in equation $\left\langle -\frac{Z}{r_1} \right\rangle = \left\langle -\frac{Z}{r_2} \right\rangle = -Z\alpha, \left\langle \frac{1}{r_{12}} \right\rangle = \frac{5}{8}\alpha$ gives the variational energy,

$$\bar{E}(\alpha) = \alpha^2 - 2Z\alpha + \frac{5}{8}\alpha$$

This will be always be an upper bound for the true ground-state energy. We can optimize our result by finding the value of α which *minimizes* the energy. We find,

$$\frac{d\bar{E}}{d\alpha} = 2\alpha - 2Z + \frac{5}{8} = 0$$

giving the optimal value.

$$\alpha = Z - \frac{5}{16}$$

This can be given a physical interpretation, noting that the parameter αα in the wavefunction represents an *effective* nuclear charge. Each electron partially shields the other electron from the positively-charged nucleus by an amount equivalent to 5/8 of an electron charge.

Substituting equation $\alpha = Z - \frac{5}{16}$ into equation $\bar{E}(\alpha) = \alpha^2 - 2Z\alpha + \frac{5}{8}\alpha$, we obtain the optimized approximation to the energy,

$$\bar{E} = -\left(Z - \frac{5}{16}\right)^2$$

For helium (Z = 2), this gives −2.84765 hartrees, an error of about 2% (E_0 = −2.90372). Note that the inequality $\bar{E} > E_0$ applies in an *algebraic* sense.

In the late 1920's, it was considered important to determine whether the helium computation could be improved, as a test of the validity of quantum mechanics for many electron systems. The table below gives the results for a selection of variational computations on helium.

Wavefunction	Parameters	Energy
$e^{-Z(r_1+r_2)}$	Z = 2	−2.75
$e^{-\alpha(r_1+r_2)}$	α = 1.6875	−2.84765
$\psi(r_1)\psi(r_2)$	best ψ(r)	−2.86168
$e^{-\alpha(r_1+r_2)}(1+cr_{12})$	best α, c	−2.89112
Hylleraas	10 parameters	−2.90363
Pekeris	1078 parameters	−2.90372

The third entry refers to the self-consistent field method, developed by Hartree. Even for the best possible choice of one-electron functions ψ(r), there remains a considerable error. This is due to failure to include the variable r_{12} in the wavefunction. The effect is known as electron correlation.

The fourth entry, containing a simple correction for correlation, gives a considerable improvement. Hylleraas extended this approach with a variational function of the form,

$$\psi(r_1, r_2, r_{12}) = e^{-\alpha(r_1+r_2)} \times \text{polynomial in } r_1, r_2, r_{12}$$

and obtained the nearly exact result with 10 optimized parameters. More recently, using modern computers, results in essentially perfect agreement with experiment have been obtained.

Spinorbitals and the Exclusion Principle

The simpler wavefunctions for helium atom in equation $\bar{\psi}(r_1, r_2) = e^{-\alpha(r_1+r_2)}$, can be interpreted as representing two electrons in hydrogen-like 1s orbitals, designated as a $1s^2$ configuration. According to Pauli's exclusion principle, which states that no two electrons in an atom can have the same set of four quantum numbers, the two 1s electrons must have *different* spins, one spin-up or α, the other spin-down or β. A product of an orbital with a spin function is called a *spinorbital*. For example, electron 1 might occupy a spinorbital which we designate,

$$\phi(1) = \psi_{1s}(1)\alpha(1) \text{ or } \psi_{1s}(1)\beta(1)$$

Spinorbitals can be designated by a single subscript, for example, ϕ_a or ϕ_b where the subscript stands for a *set* of four quantum numbers. In a two electron system the occupied spinorbitals ϕ_a and ϕ_b must be different, meaning that at least one of their four quantum numbers must be unequal. A two-electron spinorbital function of the form,

$$\Psi(1,2) = \frac{1}{2}\left(\phi_a(1)\phi_b(2) - \phi_b(1)\phi_a(2)\right)$$

automatically fulfils the Pauli principle since it vanishes if a=b. Moreover, this function associates each electron equally with each orbital, which is consistent with the *indistinguishability* of identical particles in quantum mechanics. The factor $1/\sqrt{2}$ normalizes the two-particle wavefunction, assuming that ϕ_a and ϕ_b are normalized and mutually orthogonal. The function is *antisymmetric* with respect to interchange of electron labels, meaning that,

$$\Psi(2,1) = -\Psi(1,2)$$

This antisymmetry property is an elegant way of expressing the Pauli principle.

We note, for future reference, that the function in equation $\Psi(1,2) = \frac{1}{2}\left(\phi_a(1)\phi_b(2) - \phi_b(1)\phi_a(2)\right)$ can be expressed as a 2 × 2 determinant:

$$\Psi(1,2) = \frac{1}{\sqrt{2}}\begin{vmatrix} \phi_a(1) & \phi_b(1) \\ \phi_a(2) & \phi_b(2) \end{vmatrix}$$

For the $1s^2$ configuration of helium, the two orbital functions are the same and equation $\Psi(1,2) = \frac{1}{2}\left(\phi_a(1)\phi_b(2) - \phi_b(1)\phi_a(2)\right)$ can be written,

$$\Psi(1,2) = \psi_{1s}(1)\psi_{1s}(2) \times \frac{1}{\sqrt{2}}\left(\alpha(1)\beta(2) - \beta(1)\alpha(2)\right)$$

For two-electron systems (but not for three or more electrons), the wavefunction can be factored into an orbital function times a spin function. The two-electron spin function,

$$\sigma_{0,0}(1,2) = \frac{1}{\sqrt{2}}\left(\alpha(1)\beta(2) - \beta(1)\alpha(2)\right)$$

represents the two electron spins in opposing directions (antiparallel) with a total spin angular momentum of zero. The two subscripts are the quantum numbers S and M_S for the total electron spin. Eqution $\Psi(1,2) = \psi_{1s}(1)\psi_{1s}(2) \times \frac{1}{\sqrt{2}}\left(\alpha(1)\beta(2) - \beta(1)\alpha(2)\right)$ is called the *singlet* spin state since there is only a single orientation for a total spin quantum number of zero. It is also possible to have both spins in the *same* state, provided the orbitals are different. There are three possible states for two parallel spins:

$$\sigma_{1,1}(1,2) = \alpha(1)\alpha(2)$$

$$\sigma_{1,0}(1,2) = \frac{1}{\sqrt{2}}\left(\alpha(1)\beta(2) + \beta(2)\alpha(2)\right)$$

$$\sigma_{1,-1}(1,2) = \beta(1)\beta(2)$$

These make up the *triplet* spin states, which have the three possible orientations of a total angular momentum of 1.

Excited States of Helium

The lowest excitated state of helium is represented by the electron configuration 1*s* 2*s*. The 1*s* 2*p* configuration has higher energy, even though the 2*s* and 2*p* orbitals in hydrogen are degenerate, because the 2*s* penetrates closer to the nucleus, where the potential energy is more negative. When electrons are in different orbitals, their spins can be either parallel or antiparallel. In order that the wavefunction satisfy the antisymmetry requirement, the two-electron orbital and spin functions must have *opposite* behavior under exchange of electron labels. There are four possible states from the 1*s* 2*s* configuration: a singlet state,

$$\Psi^{+}(1,2) = \frac{1}{\sqrt{2}}\left(\psi_{1s}(1)\psi_{2s}(2) + \psi_{2s}(1)\psi_{1s}(2)\right)\sigma_{0,0}(1,2)$$

and three triplet states,

$$\Psi^{-}(1,2) = \frac{1}{\sqrt{2}}\left(\psi_{1s}(1)\psi_{2s}(2) - \psi_{2s}(1)\psi_{1s}(2)\right)\begin{cases}\sigma_{1,1}(1,2)\\ \sigma_{1,0}(1,2)\\ \sigma_{1,-1}(1,2)\end{cases}$$

Using the Hamiltonian in equation $\left\{-\frac{1}{2}\nabla_1^2 - \frac{1}{2}\nabla_2^2 - \frac{Z}{r^2} + \frac{1}{r_{12}}\right\}\psi(r_1,r_2) = E\psi(r_1,r_2)$, we can compute the approximate energies,

$$E^{\pm} = \iint \Psi^{\pm}(1,2)\hat{H}\Psi^{\pm}(1,2)d\tau_1 d\tau_2$$

After evaluating some fierce-looking integrals, this reduces to the form,

$$E^{\pm} = I(1s) + I(2s) + J(1s,2s) \pm K(1s,2s)$$

in terms of the one electron integrals,

$$I(a) = \int \psi_a(r)\left\{-\frac{1}{2}\nabla^2 - \frac{Z}{r}\right\}\psi_a(r)d\tau$$

the Coulomb integrals,

$$J(a,b) = \int\int \psi_a(r_1)^2 \frac{1}{r_{12}} \psi_b(r_2)^2 d\tau_1 d\tau_2$$

and the exchange integrals,

$$K(a,b) = \int\int \psi_a(r_1)\psi_b(r_1) \frac{1}{r_{12}} \psi_a(r_2)\psi_b(r_2) d\tau_1 d\tau_2$$

The Coulomb integral represents the repulsive potential energy for two interacting charge distributions $\psi_a(r_1)^2$ and $\psi_b(r_2)^2$. The exchange integral, which has no classical analog, arises because of the exchange symmetry (or antisymmetry) requirement of the wavefunction. Both *J* and *K* can be shown to be positive quantities. Therefore the lower sign in equation $\sigma_{1,-1}(1,2) = \beta(1)\beta(2)$ represents the state of lower energy, making the triplet state of the configuration 1s 2s lower in energy than the singlet state. This is an almost universal generalization and contributes to Hund's rule.

5

Atomic Spectroscopy

The field of study which focuses on the electromagnetic radiation which is absorbed and emitted by atoms is called atomic spectroscopy. Some of the focus areas of this field are Einstein coefficients and interaction of an atom with radiation. The diverse aspects of atomic spectroscopy such as atomic spectra and alkali spectra have been thoroughly discussed in this chapter.

Atomic spectroscopy is the determination of elemental composition by its electromagnetic or mass spectrum. The study of the electromagnetic spectrum of elements is called Optical Atomic Spectroscopy. Electrons exist in energy levels within an atom. These levels have well defined energies and electrons moving between them must absorb or emit energy equal to the difference between them. In optical spectroscopy, the energy absorbed to move an electron to a more energetic level and the energy emitted as the electron moves to a less energetic energy level is in the form of a photon. The wavelength of the emitted radiant energy is directly related to the electronic transition which has occurred. Since every element has a unique electronic structure, the wavelength of light emitted is a unique property of each individual element. As the orbital configuration of a large atom may be complex, there are many electronic transitions which can occur, each transition resulting in the emission of a characteristic wavelength of light, as illustrated below.

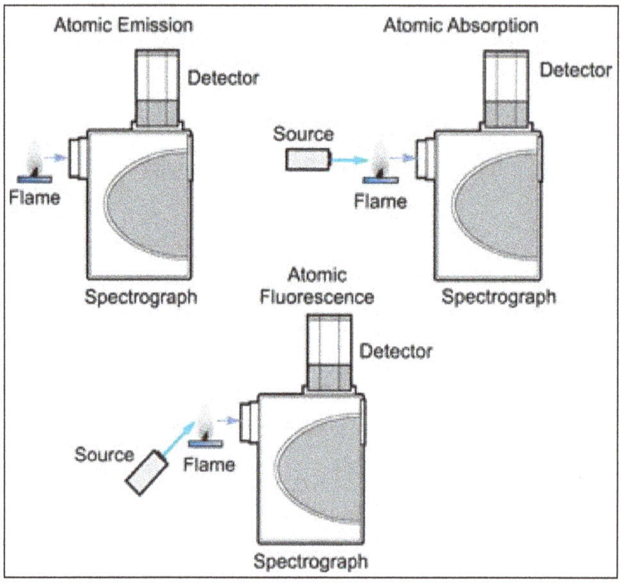

The science of atomic spectroscopy has yielded three techniques for analytical use: Atomic Absorption, Atomic Emission and Atomic Fluorescence. The process of excitation and decay to the ground state is involved in all three fields of atomic spectroscopy. Either the energy absorbed in the excitation process, or the energy emitted in the decay process is measured and used for analytical purposes.

If light of just the right wavelength impinges on a free, ground state atom, the atom may absorb the light as it enters an excited state in a process known as atomic absorption. Atomic absorption measures the amount of light at the resonant wavelength which is absorbed as it passes through a cloud of atoms. As the number of atoms in the light path increases, the amount of light absorbed increases in a predictable way. By measuring the amount of light absorbed, a quantitative determination of the amount of analyze element present can be made. The use of special light sources and careful selection of wavelength allow the specific quantitative determination of individual elements in the presence of others. The atom cloud required for atomic absorption measurements is produced by supplying enough thermal energy to the sample to dissociate the chemical compounds into free atoms. Aspirating a solution of the sample into a flame aligned in the light beam serves this purpose. Under the proper flame conditions, most of the atoms will remain in the ground state form and are capable of absorbing light at the analytical wavelength from a source lamp. The ease and speed at which precise and accurate determinations can be made with this technique have made atomic absorption one of the most popular methods for the determination of metals.

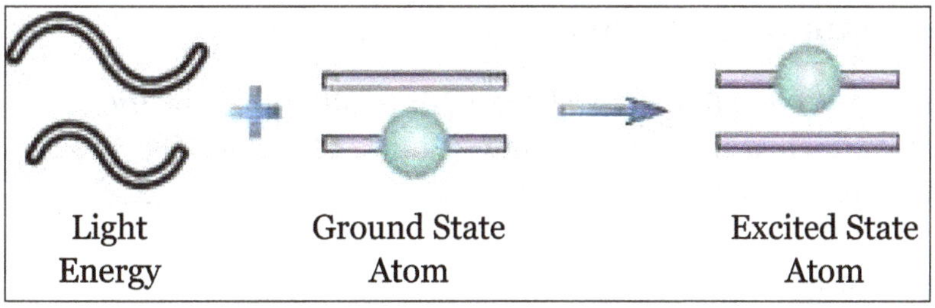

The atomic absorption process.

In atomic emission, a sample is subjected to a high energy, thermal environment in order to produce excited state atoms, capable of emitting light. The energy source can be an electrical arc, a flame, or more recently, a plasma. The emission spectrum of an element exposed to such an energy source consists of a collection of the allowable emission wavelengths, commonly called emission lines, because of the discrete nature of the emitted wavelengths. This emission spectrum can be used as a unique characteristic for qualitative identification of the element. Atomic emission using electrical arcs has been widely used in qualitative analysis. Emission techniques can also be used to determine how much of an element is present in a sample. For a "quantitative" analysis, the intensity of light emitted at the wavelength of the element to be determined is measured. The emission intensity at this wavelength will be greater as the number of atoms of the analyte element increases. The technique of flame photometry is an application of atomic emission for quantitative analysis.

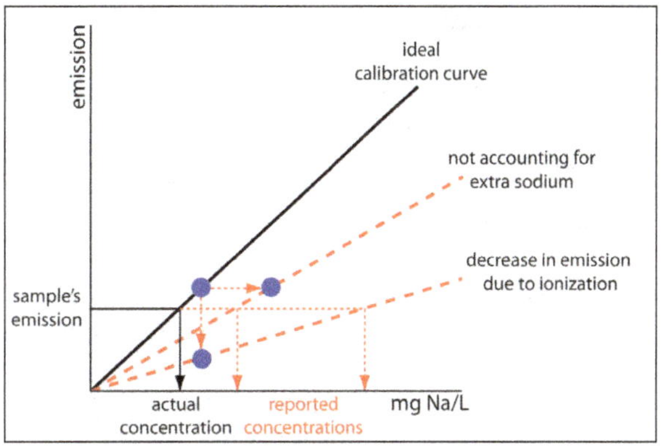

Atomic spectroscopy.

The third field of atomic spectroscopy is atomic fluorescence. This technique incorporates aspects of both atomic absorption and atomic emission. Like atomic absorption, ground state atoms created in a flame are excited by focusing a beam of light into the atomic vapor. Instead of looking at the amount of light absorbed in the process, however, the emission resulting from the decay of the atoms excited by the source light is measured. The intensity of this "fluorescence" increases with increasing atom concentration, providing the basis for quantitative determination. The source lamp for atomic fluorescence is mounted at an angle to the rest of the optical system, so that the light detector sees only the fluorescence in the flame and not the light from the lamp itself. It is advantageous to maximize lamp intensity since sensitivity is directly related to the number of excited atoms which in turn is a function of the intensity of the exciting radiation.

While atomic absorption is the most widely applied of the three techniques and usually offers several advantages over the other two, particular benefits may be gained with either emission or fluorescence in special analytical situations.

Commonly used Atomic Spectroscopy Techniques

There are three widely accepted analytical methods – atomic absorption, atomic emission and mass spectrometry – which will form the focus of our discussion, allowing us to go into greater depth on the most common techniques in use today.

Atomic Absorption Spectroscopy

Atomic-absorption (AA) spectroscopy uses the absorption of light to measure the concentration of gas-phase atoms. Since samples are usually liquids or solids, the analyte atoms or ions must be vaporized in a flame or graphite furnace.

The atoms absorb ultraviolet or visible light and make transitions to higher electronic energy levels. The analyte concentration is determined from the amount of absorption.

Applying the Beer-Lambert law directly in AAS is difficult due to variations in the atomization efficiency from the sample matrix, and nonuniformity of concentration and path length of analyte atoms (in graphite furnace AA). Concentration measurements are usually determined from a working curve after calibrating the instrument with standards of known concentration.

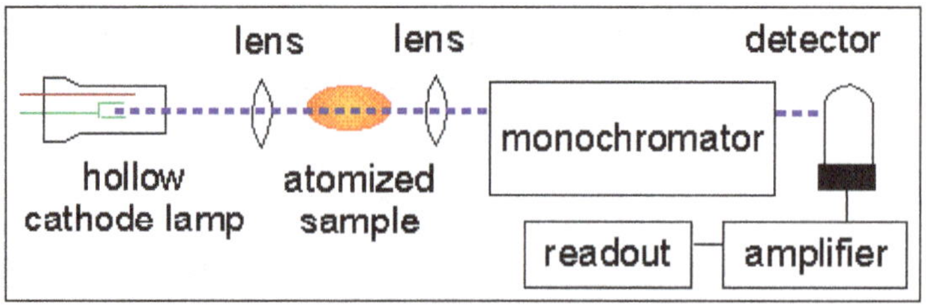

Schematic of an Atomic-absorption Experiment.

Any atom has its own distinct pattern of wavelengths at which it will absorb energy, due to the unique configuration of electrons in its outer shell. This allows for the qualitative analysis of a pure sample.

In order to tell how much of a known element is present in a sample, one must first establish a basis for comparison using known quantities. It can be done producing a calibration curve. For this process, a known wavelength is selected, and the detector will measure only the energy emitted at that wavelength. However, as the concentration of the target atom in the sample increases, absorption will also increase proportionally. Thus, one runs a series of known concentrations of some compound, and records the corresponding degree of absorbance, which is an inverse percentage of light transmitted. A straight line can then be drawn between all of the known points. From this line, one can then extrapolate the concentration of the substance under investigation from its absorbance. The use of special light sources and specific wavelength selection allows the quantitative determination of individual components of a multielement mixture.

The phenomenon of atomic absorption (AA) was first observed in 1802 with the discovery of the Fraunhofer lines in the sun's spectrum. It was not until 1953 that Australian physicist Sir Alan Walsh demonstrated that atomic absorption could be used as a quantitative analtical tool. Atomic absorption analysis involves measuring the absorption of light by vaporized ground state atoms and relating the absorption to concentration. The incident light beam is attenuated by atomic vapor absorption according to Beer's law.

The process of atomic absorption spectroscopy (AAS) involves two steps:

1. Atomization of the sample.
2. The absorption of radiation from a light source by the free atoms.

The sample, either a liquid or a solid, is atomized in either a flame or a graphite furnace. Upon the absorption of ultraviolet or visible light, the free atoms undergo electronic transitions from the ground state to excited electronic states.

To obtain the best results in AA, the instrumental and chemical parameters of the system must be geared toward the production of neutral ground state atoms of the element of interest. A common method is to introduce a liquid sample into a flame. Upon introduction, the sample solution is dispersed into a fine spray, the spray is then desolvated into salt particles in the flame and the particles are subsequently vaporized into neutral atoms, ionic species and molecular species. All of these conversion processes occur in geometrically definable regions in the flame. It is therefore important to set the instrument parameters such that the light from the source (typically a hollow-cathode lamp) is directed through the region of the flame that contains the maximum number of neutral atoms. The light produced by the hollow-cathode lamp is emitted from excited atoms of the same element which is to be determined. Therefore the radiant energy corresponds directly to the wavelength which is absorbable by the atomized sample. This method provides both sensitivity and selectivity since other elements in the sample will not generally absorb the chosen wavelength and thus, will not interfere with the measurement. To reduce background interference, the wavelength of interest is isolated by a monochromator placed between the sample and the detector.

Instrumentation

Atomic absorption spectrophotometers use the same single-beam or double-beam optics described earlier for molecular absorption spectrophotometers. There is, however, an important additional need in atomic absorption spectroscopy—we must covert the analyte into free atoms. In most cases our analyte is in solution form. If our sample is a solid, then we must bring it into solution before the analysis. When analyzing a lake sediment for Cu, Zn, and Fe, for example, we bring the analytes into solution as Cu^{2+}, Zn^{2+}, and Fe^{3+} by extracting them with a suitable reagent.

Atomization

The process of converting an analyte to a free gaseous atom is called atomization. Converting an aqueous analyte into a free atom requires that we strip away the solvent, volatilize the analytes, and, if necessary, dissociate the analyte into free atoms. Desolvating an aqueous solution of $CuCl_2$, for example, leaves us with solid particulates of $CuCl_2$. Converting the particulate $CuCl_2$ to gas phases atoms of Cu and Cl requires thermal energy.

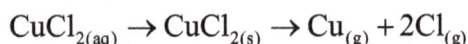

$$CuCl_{2(aq)} \rightarrow CuCl_{2(s)} \rightarrow Cu_{(g)} + 2Cl_{(g)}$$

There are two common atomization methods: flame atomization and electrothermal atomization, although a few elements are atomized using other methods.

Flame Atomizer

Figure shows a typical flame atomization assembly with close-up views of several key components. In the unit shown here, the aqueous sample is drawn into the assembly by passing a high-pressure stream of compressed air past the end of a capillary tube immersed in the sample. When the sample exits the nebulizer it strikes a glass impact bead, converting it into a fine aerosol mist within the spray chamber. The aerosol mist is swept through the spray chamber by the combustion gases—compressed air and acetylene in this case–to the burner head where the flame's thermal energy desolvates the aerosol mist to a dry aerosol of small, solid particles. The flame's thermal energy then volatilizes the particles, producing a vapor consisting of molecular species, ionic species, and free atoms.

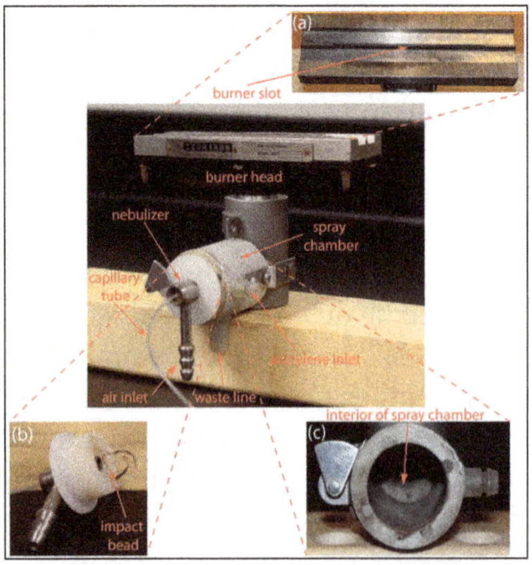

Flame atomization assembly with expanded views of: (a) the burner head showing the burner slot where the flame is located; (b) the nebulizer's impact bead; and (c) the interior of the spray chamber. Although the unit shown here is from an older instrument, the basic components of a modern flame AA spectrometer are the same.

1. Burner: The slot burner in figure provides a long optical path length and a stable flame. Because absorbance increases linearly with the path length, a long path length provides greater sensitivity. A stable flame minimizes uncertainty due to fluctuations in the flame.

The burner is mounted on an adjustable stage that allows the entire assembly to move horizontally and vertically. Horizontal adjustments ensure that the flame is aligned with the instrument's optical path. Vertical adjustments adjust the height within the flame from which absorbance is monitored. This is important because two competing processes affect the concentration of free atoms in the flame. The more time the analyte spends in the flame the greater the atomization efficiency; thus, the production of free atoms increases with height. On the other hand, a longer residence time allows more opportunity for the free atoms to combine with oxygen to form a molecular oxide. For an easily oxidized metal, such as Cr, the concentration of free atoms is greatest just above the burner head. For metals, such as Ag, which are difficult to oxidize, the concentration of free atoms increases steadily with height. Other atoms show concentration profiles that maximize at a characteristic height.

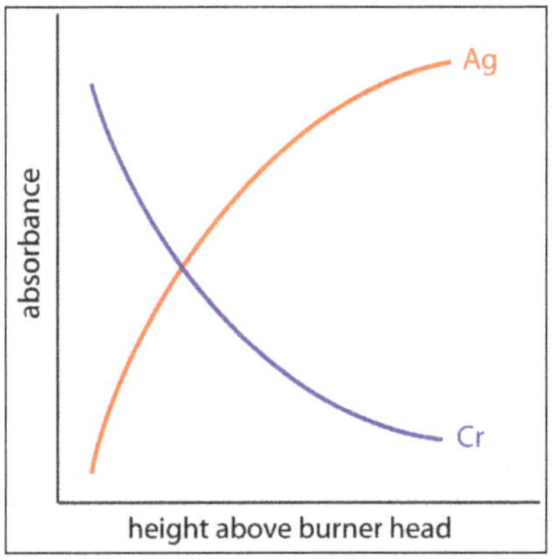

Absorbance versus height profiles for Ag and Cr in flame atomic absorption spectroscopy.

2. Flame: The flame's temperature, which affects the efficiency of atomization, depends on the fuel–oxidant mixture, several examples of which are listed in table. Of these, the air–acetylene and the nitrous oxide–acetylene flames are the most popular. Normally the fuel and oxidant are mixed in an approximately stoichiometric ratio; however, a fuel-rich mixture may be necessary for easily oxidized analytes.

Table: Fuels and oxidants used for flame combustion.

Fuel	Oxidant	Temperature range (°c)
Natural gas	Air	1700–1900
Hydrogen	Air	2000–2100
Acetylene	Air	2100–2400
Acetylene	Nitrous oxide	2600–2800
Acetylene	Oxygen	3050–3150

Figure shows a cross-section through the flame, looking down the source radiation's optical path. The primary combustion zone is usually rich in gas combustion products that emit radiation, limiting is usefulness for atomic absorption. The interzonal region generally is rich in free atoms and provides the best location for measuring atomic absorption. The hottest part of the flame is typically 2–3 cm above the primary combustion zone. As atoms approach the flame's secondary combustion zone, the decrease in temperature allows for formation of stable molecular species.

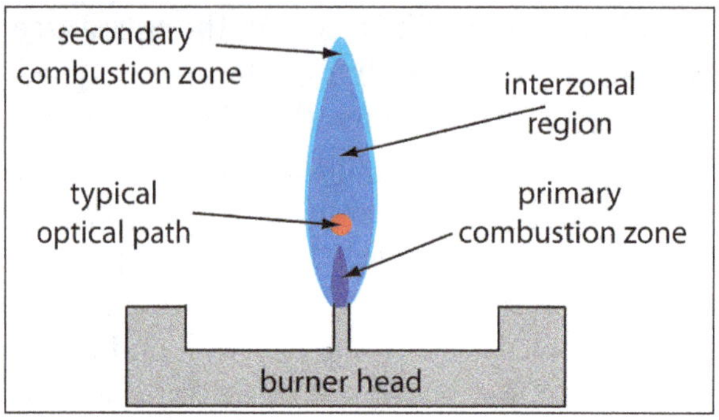

Profile of typical flame using a slot burner. The relative size of each zone depends on many factors, including the choice of fuel and oxidant, and their relative proportions.

3. Sample Introduction: The most common means for introducing samples into a flame atomizer is a continuous aspiration in which the sample flows through the burner while we monitor the absorbance. Continuous aspiration is sample intensive, typically requiring from 2–5 mL of sample.

Flame microsampling allows us to introduce a discrete sample of fixed volume, and is useful when we have a limited amount of sample or when the sample's matrix is incompatible with the flame atomizer. For example, continuously aspirating a sample that has a high concentration of dissolved solids—sea water, for example, comes to mind—may build-up a solid deposit on the burner head that obstructs the flame and that lowers the absorbance. Flame microsampling is accomplished using a micropipet to place 50–250 μL of sample in a Teflon funnel connected to the nebulizer, or by dipping the nebulizer tubing into the sample for a short time. Dip sampling is usually accomplished with an automatic sampler. The signal for flame microsampling is a transitory peak whose height or area is proportional to the amount of analyte that is injected.

4. Advantages and Disadvantages of Flame Atomization: The principal advantage of flame atomization is the reproducibility with which the sample is introduced into the spectrophotometer. A significant disadvantage to flame atomizers is that the efficiency of atomization may be quite poor. There are two reasons for poor atomization efficiency. First, the majority of the aerosol droplets produced during nebulization are too large to be carried to the flame by the combustion gases. Consequently, as much as 95% of the sample never reaches the flame. A second reason for poor atomization efficiency is that the large volume of combustion gases significantly dilutes the sample. Together, these contributions to the efficiency of atomization reduce sensitivity because the analyte's concentration in the flame may be a factor of 2.5×10^{-6} less than that in solution. This is the reason for the waste line shown at the bottom of the spray chamber in figure.

Electrothermal Atomizers

A significant improvement in sensitivity is achieved by using the resistive heating of a graphite tube in place of a flame. A typical electrothermal atomizer, also known as a graphite furnace, consists of a cylindrical graphite tube approximately 1–3 cm in length and 3–8 mm in diameter. As shown in figure, the graphite tube is housed in a sealed assembly that has optically transparent windows at each end. A continuous stream of an inert gas is passed through the furnace, protecting the graphite tube from oxidation and removing the gaseous products produced during atomization. A power supply is used to pass a current through the graphite tube, resulting in resistive heating.

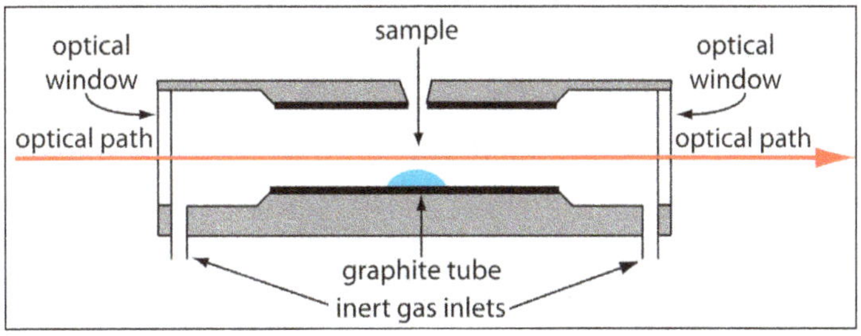

Diagram showing a cross-section of an electrothermal analyzer.

Samples of between 5–50 μL are injected into the graphite tube through a small hole at the top of the tube. Atomization is achieved in three stages. In the first stage the sample is dried to a solid residue using a current that raises the temperature of the graphite tube to about 110 °C. In the second stage, which is called ashing, the temperature is increased to between 350–1200 °C. At these temperatures any organic material in the sample is converted to CO_2 and H_2O, and volatile inorganic materials are vaporized. These gases are removed by the inert gas flow. In the final stage the sample is atomized by rapidly increasing the temperature to between 2000–3000 °C. The result is a transient absorbance peak whose height or area is proportional to the absolute amount of analyte injected into the graphite tube. Together, the three stages take approximately 45–90 s, with most of this time used for drying and ashing the sample.

Electrothermal atomization provides a significant improvement in sensitivity by trapping the gaseous analyte in the small volume within the graphite tube. The analyte's concentration in the resulting vapor phase may be as much as 1000× greater than in a flame atomization. This improvement in sensitivity—and the resulting improvement in detection limits—is offset by a significant decrease in precision. Atomization efficiency is strongly influenced by the sample's contact with the graphite tube, which is difficult to control reproducibly.

Miscellaneous Atomization Methods

A few elements may be atomized by a chemical reaction that produces a volatile product. Elements such as As, Se, Sb, Bi, Ge, Sn, Te, and Pb, for example, form volatile hydrides when reacted with $NaBH_4$ in acid. An inert gas carries the volatile hydrides to either a flame or to a heated quartz observation tube situated in the optical path. Mercury is determined by the cold-vapor method in which it is reduced to elemental mercury with $SnCl_2$. The volatile Hg is carried by an inert gas to an unheated observation tube situated in the instrument's optical path.

Quantitative Applications

Atomic absorption is widely used for the analysis of trace metals in a variety of sample matrices. Using Zn as an example, atomic absorption methods have been developed for its determination in samples as diverse as water and wastewater, air, blood, urine, muscle tissue, hair, milk, breakfast cereals, shampoos, alloys, industrial plating baths, gasoline, oil, sediments, and rocks.

Developing a quantitative atomic absorption method requires several considerations, including choosing a method of atomization, selecting the wavelength and slit width, preparing the sample for analysis, minimizing spectral and chemical interferences, and selecting a method of standardization.

Developing a Quantitative Method

1. Flame or Electrothermal Atomization: The most important factor in choosing a method of atomization is the analyte's concentration. Because of its greater sensitivity, it takes less analyte to achieve a given absorbance when using electrothermal atomization. In table below, which compares the amount of analyte needed to achieve an absorbance of 0.20 when using flame atomization and electrothermal atomization, is useful when selecting an atomization method. For example, flame atomization is the method of choice if our samples contain 1–10 mg Zn^{2+}/L, but electrothermal atomization is the best choice for samples containing 1–10 μg Zn^{2+}/L.

Table: Concentration of analyte yielding an absorbance of 0.20.

	Concentration (mg/L)	
Element	Flame atomization	Electrothermal atomization
Ag	1.5	0.0035
Al	40	0.015
As	40	0.050
Ca	0.8	0.003
Cd	0.6	0.001
Co	2.5	0.021
Cr	2.5	0.0075
Cu	1.5	0.012
Fe	2.5	0.006
Hg	70	0.52
Mg	0.15	0.00075
Mn	1	0.003
Na	0.3	0.00023
Ni	2	0.024
Pb	5	0.080
Pt	70	0.29
Sn	50	0.023
Zn	0.3	0.00071

2. Selecting the Wavelength and Slit Width: The source for atomic absorption is a hollow cathode lamp consisting of a cathode and anode enclosed within a glass tube filled with a low pressure of Ne or Ar. Applying a potential across the electrodes ionizes the filler gas. The positively charged gas ions collide with the negatively charged cathode, sputtering atoms from the cathode's surface. Some of the sputtered atoms are in the excited state and emit radiation characteristic of the metal(s) from which the cathode was manufactured. By fashioning the cathode from the metallic analyte, a hollow cathode lamp provides emission lines that correspond to the analyte's absorption spectrum.

Each element in a hollow cathode lamp provides several atomic emission lines that we can use for atomic absorption. Usually the wavelength that provides the best sensitivity is the one we choose to use, although a less sensitive wavelength may be more appropriate for a larger concentration of analyte. For the Cr hollow cathode lamp in table, for example, the best sensitivity is obtained using a wavelength of 357.9 nm.

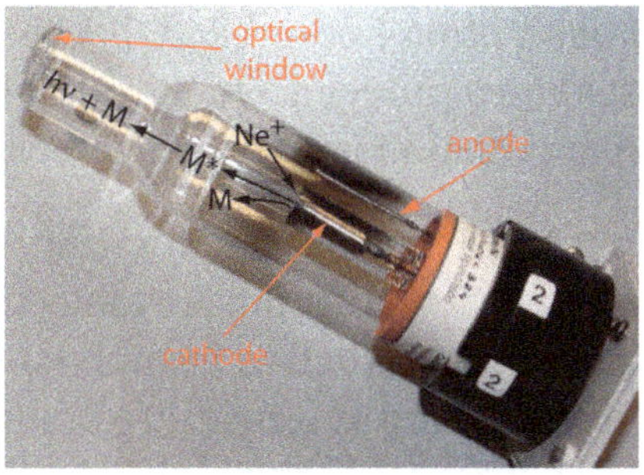

Photo of a typical multielemental hollow cathode lamp. The cathode in this lamp is fashioned from an alloy containing Co, Cr, Cu, Fe, Mn, and Ni, and is surrounded by a glass shield to isolate it from the anode. The lamp is filled with Ne gas. Also shown is the process leading to atomic emission.

Another consideration is the intensity of the emission line. If several emission lines meet our need for sensitivity, we may wish to use the emission line with the largest relative P_0 because there is less uncertainty in measuring P_0 and P_T. When analyzing samples containing ≈10 mg Cr/L, for example, the first three wavelengths in table provide an appropriate sensitivity. The wavelengths of 425.5 nm and 429.0 nm, however, have a greater P_0 and will provide less uncertainty in the measured absorbance.

Table: Atomic emission lines for a Cr hollow cathode lamp.

Wavelength (nm)	Slit width (nm)	Mg Cr/l giving A = 0.20	P_0 (relative)
357.9	0.2	2.5	40
425.4	0.2	12	85
429.0	0.5	20	100
520.5	0.2	1500	15
520.8	0.2	500	20

The emission spectrum from a hollow cathode lamp includes, besides emission lines for the analyte, additional emission lines for impurities present in the metallic cathode and from the filler gas. These additional lines are a source of stray radiation that leads to an instrumental deviation from Beer's law. The monochromator's slit width is set as wide as possible, improving the throughput of radiation, while, at the same time, being narrow enough to eliminate the stray radiation.

1. Preparing the Sample: Flame and electrothermal atomization require that the sample be in solution. Solid samples are brought into solution by dissolving in an appropriate solvent. If the sample is not soluble it may be digested, either on a hot-plate or by microwave, using HNO_3, H_2SO_4, or $HClO_4$. Alternatively, we can extract the analyte using a Soxhlet extractor. Liquid samples may be analyzed directly or extracted if the matrix is incompatible with the method of atomization. A serum sample, for instance, is difficult to aspirate when using flame atomization and may produce an unacceptably high background absorbance when using electrothermal atomization. A liquid–liquid extraction using an organic solvent and a chelating agent is frequently used to concentrate analytes. Dilute solutions of Cd^{2+}, Co^{2+}, Cu^{2+}, Fe^{3+}, Pb^{2+}, Ni^{2+}, and Zn^{2+}, for example, can be

concentrated by extracting with a solution of ammonium pyrrolidine dithiocarbamate in methyl isobutyl ketone.

2. Minimizing Spectral Interference: A spectral interference occurs when an analyte's absorption line overlaps with an interferent's absorption line or band. Because they are so narrow, the overlap of two atomic absorption lines is seldom a problem. On the other hand, a molecule's broad absorption band or the scattering of source radiation is a potentially serious spectral interference.

An important consideration when using a flame as an atomization source is its effect on the measured absorbance. Among the products of combustion are molecular species that exhibit broad absorption bands and particulates that scatter radiation from the source. If we fail to compensate for these spectral interference, then the intensity of transmitted radiation decreases. The result is an apparent increase in the sample's absorbance. Fortunately, absorption and scattering of radiation by the flame are corrected by analyzing a blank.

Spectral interferences also occur when components of the sample's matrix other than the analyte react to form molecular species, such as oxides and hydroxides. The resulting absorption and scattering constitutes the sample's background and may present a significant problem, particularly at wavelengths below 300 nm where the scattering of radiation becomes more important. If we know the composition of the sample's matrix, then we can prepare our samples using an identical matrix. In this case the background absorption is the same for both the samples and standards. Alternatively, if the background is due to a known matrix component, then we can add that component in excess to all samples and standards so that the contribution of the naturally occurring interferent is insignificant. Finally, many interferences due to the sample's matrix can be eliminated by increasing the atomization temperature. For example, by switching to a higher temperature flame it may be possible to prevent the formation of interfering oxides and hydroxides.

If the identity of the matrix interference is unknown, or if it is not possible to adjust the flame or furnace conditions to eliminate the interference, then we must find another method to compensate for the background interference. Several methods have been developed to compensate for matrix interferences, and most atomic absorption spectrophotometers include one or more of these methods.

One of the most common methods for background correction is to use a continuum source, such as a D_2 lamp. Because a D_2 lamp is a continuum source, absorbance of its radiation by the analyte's narrow absorption line is negligible. Only the background, therefore, absorbs radiation from the D_2 lamp. Both the analyte and the background, on the other hand, absorb the hollow cathode's radiation. Subtracting the absorbance for the D_2 lamp from that for the hollow cathode lamp gives a corrected absorbance that compensates for the background interference. Although this method of background correction may be quite effective, it does assume that the background absorbance is constant over the range of wavelengths passed by the monochromator. If this is not true, subtracting the two absorbances may underestimate or overestimate the background.

3. Minimizing Chemical Interferences: The quantitative analysis of some elements is complicated by chemical interferences occurring during atomization. The two most common chemical interferences are the formation of nonvolatile compounds containing the analyte and ionization of the analyte.

One example of the formation of nonvolatile compounds is the effect of PO_4^{3-} or Al^{3+} on the flame atomic absorption analysis of Ca^{2+}. In one study, for example, adding 100 ppm Al^{3+} to a solution of 5 ppm Ca^{2+} decreased the calcium ion's absorbance from 0.50 to 0.14, while adding 500 ppm PO_4^{3-} to a similar solution of Ca^{2+} decreased the absorbance from 0.50 to 0.38. These interferences were attributed to the formation of nonvolatile particles of $Ca_3(PO_4)_2$ and an Al–Ca–O oxide.

When using flame atomization, we can minimize the formation of nonvolatile compounds by increasing the flame's temperature, either by changing the fuel-to-oxidant ratio or by switching to a different combination of fuel and oxidant. Another approach is to add a releasing agent or a protecting agent to the samples. A releasing agent is a species that reacts with the interferent, releasing the analyte during atomization. Adding Sr^{2+} or La^{3+} to solutions of Ca^{2+}, for example, minimizes the effect of PO_4^{3-} and Al^{3+} by reacting in place of the analyte. Thus, adding 2000 ppm $SrCl_2$ to the Ca^{2+}/PO_4^{3-} and Ca^{2+}/Al^{3+} mixtures described in the previous paragraph increased the absorbance to 0.48. A protecting agent reacts with the analyte to form a stable volatile complex. Adding 1% w/w EDTA to the Ca^{2+}/PO_4^{3-} solution described in the previous paragraph increased the absorbance to 0.52.

Ionization interferences occur when thermal energy from the flame or the electrothermal atomizer is sufficient to ionize the analyte,

$$M_{(g)} \overset{\Delta}{\rightleftharpoons} M^{+}_{(g)} + e^{-}$$

where M is the analyte. Because the absorption spectra for M and M^+ are different, the position of the equilibrium in reaction $M_{(g)} \overset{\Delta}{\rightleftharpoons} M^{+}_{(g)} + e^{-}$ affects absorbance at wavelengths where M absorbs. To limit ionization we add a high concentration of an ionization suppressor, which is simply a species that ionizes more easily than the analyte. If the concentration of the ionization suppressor is sufficient, then the increased concentration of electrons in the flame pushes reaction $M_{(g)} \overset{\Delta}{\rightleftharpoons} M^{+}_{(g)} + e^{-}$, preventing the analyte's ionization. Potassium and cesium are frequently used as an ionization suppressor because of their low ionization energy.

4. Standardizing the Method: Because Beer's law also applies to atomic absorption, we might expect atomic absorption calibration curves to be linear. In practice, however, most atomic absorption calibration curves are nonlinear, or linear for only a limited range of concentrations. Nonlinearity in atomic absorption is a consequence of instrumental limitations, including stray radiation from the hollow cathode lamp and the variation in molar absorptivity across the absorption line. Accurate quantitative work, therefore, often requires a suitable means for computing the calibration curve from a set of standards.

When possible, a quantitative analysis is best conducted using external standards. Unfortunately, matrix interferences are a frequent problem, particularly when using electrothermal atomization. For this reason the method of standard additions is often used. One limitation to this method of standardization, however, is the requirement that there be a linear relationship between absorbance and concentration.

Evaluation of Atomic Absorption Spectroscopy

Scale of Operation

Atomic absorption spectroscopy is ideally suited for the analysis of trace and ultratrace analytes, particularly when using electrothermal atomization. For minor and major analyte, sample can be diluted before the analysis. Most analyses use a macro or a meso sample. The small volume requirement for electrothermal atomization or flame microsampling however, makes practical the analysis micro and ultramicro samples.

Accuracy

If spectral and chemical interferences are minimized, an accuracy of 0.5–5% is routinely attainable. When the calibration curve is nonlinear, accuracy may be improved by using a pair of standards whose absorbances closely bracket the sample's absorbance and assuming that the change in absorbance is linear over this limited concentration range. Determinate errors for electrothermal atomization are often greater than that obtained with flame atomization due to more serious matrix interferences.

Precision

For absorbance values greater than 0.1–0.2, the relative standard deviation for atomic absorption is 0.3–1% for flame atomization and 1–5% for electrothermal atomization. The principle limitation is the variation in the concentration of free analyte atoms resulting from variations in the rate of aspiration, nebulization, and atomization when using a flame atomizer, and the consistency of injecting samples when using electrothermal atomization.

Sensitivity

The sensitivity of a flame atomic absorption analysis is influenced strongly by the flame's composition and by the position in the flame from which we monitor the absorbance. Normally the sensitivity of an analysis is optimized by aspirating a standard solution of the analyte and adjusting operating conditions, such as the fuel-to-oxidant ratio, the nebulizer flow rate, and the height of the burner, to give the greatest absorbance. With electrothermal atomization, sensitivity is influenced by the drying and ashing stages that precede atomization. The temperature and time used for each stage must be optimized for each type of sample.

Sensitivity is also influenced by the sample's matrix. For example, that sensitivity can be decreased by chemical interferences. An increase in sensitivity may be realized by adding a low molecular weight alcohol, ester, or ketone to the solution, or by using an organic solvent.

Selectivity

Due to the narrow width of absorption lines, atomic absorption provides excellent selectivity. Atomic absorption can be used for the analysis of over 60 elements at concentrations at or below the level of μg/L.

Time, Cost and Equipment

The analysis time when using flame atomization is short, with sample throughputs of 250–350 determinations per hour when using a fully automated system. Electrothermal atomization requires substantially more time per analysis, with maximum sample throughputs of 20–30 determinations per hour. The cost of a new instrument ranges from between $10,000–$50,000 for flame atomization, and from $18,000–$70,000 for electrothermal atomization. The more expensive instruments in each price range include double-beam optics, automatic samplers, and can be programmed for multielemental analysis by allowing the wavelength and hollow cathode lamp to be changed automatically.

Techniques of Measurement and EPA Methods using FAAS

Atomic absorption spectrometry is a fairly universal analytical method for determination of metallic elements when present in both trace and major concentrations. The EPA employs this technique for determining the metal concentration in samples from a variety of matrices.

A) Sample preparation: Depending on the information required, total recoverable metals, dissolved metals, suspended metals, and total metals could be obtained from a certain environmental matrix. Table below lists the EPA method number for sample processing in terms of the environmental matrices and information required.

Table: EPA sample processing method for metallic element analysis.

Analysis Target	Method Number	Environmental Matrice
Total recoverable metals	3005	Ground water/surface water
Dissolved metals	3005	Ground water/surface water
Suspended metals	3005	Ground water/surface water
Total metals	3010	Aqueous samples, wastes that contain suspended solids and mobility-procedure extracts
Total metals	3015	Aqueous samples, wastes that contain suspended solids and mobility-procedure extracts
Total metals	3020	Aqueous samples, wastes that contain suspended solids and mobility-procedure extracts
Total metals	3050	Sediments, sludges and soil samples
Total metals	3051	Sludges, sediment, soil and oil

Appropriate acid digestion is employed in these methods. Hydrochloric acid digestion is not suitable for samples, which will be analyzed by graphite furnace atomic absorption spectroscopy because it can cause interferences during furnace atomization.

B) Calibration and standard curves: As with other analytical techniques, atomic absorption spectrometry requires careful calibration. EPA QA/QC demands calibration through several steps including interference check sample, calibration verification, calibration standards, bland control, and linear dynamic range.

The idealized calibration or standard curve is stated by Beer's law that the absorbance of an absorbing analyte is proportional to its concentration.

Unfortunately, deviations from linearity usually occur, especially as the concentration of metallic analytes increases due to various reasons, such as unabsorbed radiation, stray light, or disproportionate decomposition of molecules at high concentrations. Figure shows an idealized and deviation of response curve. The curvature could be minimized, although it is impossible to be avoided completely. It is desirable to work in the linearity response range. The rule of thumb is that a minimum of five standards and a blank should be prepared in order to have sufficient information to fit the standard curve appropriately. Manufacturers should be consulted if a manual curvature correction function is available for a specific instrument.

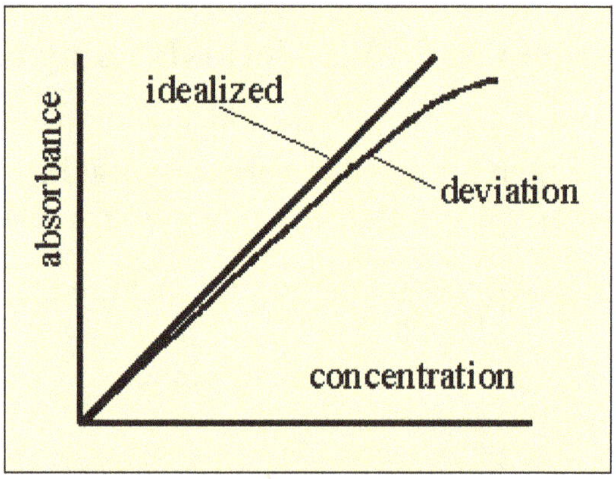

Idealized/deviation response curve.

If the sample concentration is too high to permit accurate analysis in linearity response range, there are three alternatives that may help bring the absorbance into the optimum working range:

- Sample dilution.
- Using an alternative wavelength having a lower absorptivity.
- Reducing the path length by rotating the burner hand.

C) EPA method for metal analysis: Flame atomic absorption methods are referred to as direct aspiration determinations. They are normally completed as single element analyses and are relatively free of interelement spectral interferences. For some elements, the temperature or type of flame used is critical. If flame and analytical conditions are not properly used, chemical and ionization interferences can occur.

Graphite furnace atomic absorption spectrometry replaces the flame with an electrically heated graphite furnace. The major advantage of this technique is that the detection limit can be extremely low. It is applicable for relatively clean samples, however, interferences could be a real problem. It is important for the analyst to establish a set of analytical protocol which is appropriate for the sample to be analyzed and for the information required. Table lists the available method for different metal analysis provided in EPA manual SW-846.

Table: EPA methods for determination of metals by direct aspiration.

Analyte	Method number	Analyte	Method number	Analyte	Method number
Aluminum	7020	Antimony	7040	Barium	7080A
Beryllium	7090	Cadmium	7130	Calcium	7140
Chromium	7190	Cobalt	7200	Copper	7210
Iron	7380	Lead	7420	Lithium	7430
Magnesium	7450	Manganese	7460	Molybdenum	7480

Nickel	7520	Osmium	7550	Potassium	7610
Silver	7760a	Sodium	7770	Strontium	7780
Thallium	7840	Tin	7870	Vanadium	7910
Zinc	7951	-	-	-	-

D) Interferences: Since the concentration of the analyte element is considered to be proportional to the ground state atom population in the flame, any factor that affects the ground state population of the analyte element can be classified as interference. Factors that may affect the ability of the instrument to read this parameter can also be classified as interference. The following are the most common interferences:

- Spectral interferences are due to radiation overlapping that of the light source. The interference radiation may be an emission line of another element or compound, or general background radiation from the flame, solvent, or analytical sample. This usually occurs when using organic solvents, but can also happen when determining sodium with magnesium present, iron with copper or iron with nickel.
- Formation of compounds that do not dissociate in the flame. The most common example is the formation of calcium and strontium phosphates.
- Ionization of the analyte reduces the signal. This is commonly happens to barium, calcium, strontium, sodium and potassium.
- Matrix interferences due to differences between surface tension and viscosity of test solutions and standards.
- Broadening of a spectral line, which can occur due to a number of factors. The most common line width broadening effects are:
  - Doppler Effect: This effect arises because atoms will have different components of velocity along the line of observation.
  - Lorentz Effect: This effect occurs as a result of the concentration of foreign atoms present in the environment of the emitting or absorbing atoms. The magnitude of the broadening varies with the pressure of the foreign gases and their physical properties.
  - Quenching Effect: In a low-pressure spectral source, quenching collision can occur in flames as the result of the presence of foreign gas molecules with vibration levels very close to the excited state of the resonance line.
  - Self-absorption or Self-reversal Effect: The atoms of the same kind as that emitting radiation will absorb maximum radiation at the center of the line than at the wings, resulting in the change of shape of the line as well as its intensity. This effect becomes serious if the vapor, which is absorbing radiation is considerably cooler than that which is emitting radiation.

Flame Atomic Absorption Spectroscopy

Atomic Absorption (AA) occurs when a ground state atom absorbs energy in the form of light of a specific wavelength and is elevated to an excited state. The amount of light energy absorbed at this wavelength will increase as the number of atoms of the selected element in the light path increases. The relationship between the amount of light absorbed and the concentration of analytes present in known standards can be used to determine unknown sample concentrations by measuring the amount of light they absorb.

Performing atomic absorption spectroscopy requires a primary light source, an atom source, a monochromator to isolate the specific wavelength of light to be measured, a detector to measure the light accurately, electronics to process the data signal and a data display or reporting system to show the results. The light source normally used is a hollow cathode lamp (HCL) or an electrodeless discharge lamp (EDL). In general, a different lamp is used for each element to be determined, although in some cases, a few elements may be combined in a multi-element lamp. In the past, photomultiplier tubes have been used as the detector. However, in most modern instruments, solid-state detectors are now used. Flow Injection Mercury Systems (FIMS) are specialized, easy-to-operate atomic absorption spectrometers for the determination of mercury. These instruments use a high-performance single-beam optical system with a low-pressure mercury lamp and solar-blind detector for maximum performance.

Whatever the system, the atom source used must produce free analyte atoms from the sample. The source of energy for freeatom production is heat, most commonly in the form of an air/acetylene or nitrous-oxide/acetylene flame. The sample is introduced as an aerosol into the flame by the sample-introduction system consisting of a nebulizer and spray chamber. The burner head is aligned so that the light beam passes through the flame, where the light is absorbed.

The major limitation of Flame AA is that the burner-nebulizer system is a relatively inefficient sampling device. Only a small fraction of the sample reaches the flame, and the atomized sample passes quickly through the light path. An improved sampling device would atomize the entire sample and retain the atomized sample in the light path for an extended period of time, enhancing the sensitivity of the technique. Which leads us to the next option – electrothermal vaporization using a graphite furnace.

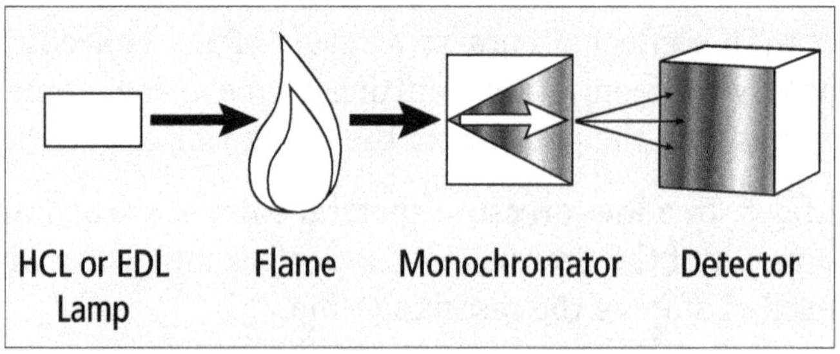

Simplified drawing of a Flame AA system.

Graphite Furnace Atomic Absorption Spectroscopy

With Graphite Furnace Atomic Absorption (GFAA), the sample is introduced directly into a graphite tube, which is then heated in a programmed series of steps to remove the solvent and major matrix components and to atomize the remaining sample. All of the analyte is atomized, and the atoms are retained within the tube (and the light path, which passes through the tube) for an extended period of time. As a result, sensitivity and detection limits are significantly improved over Flame AA.

Graphite Furnace analysis times are longer than those for Flame sampling and fewer elements can be determined using GFAA. However, the enhanced sensitivity of GFAA, and its ability to analyze very small samples, significantly expands the capabilities of atomic absorption.

GFAA allows the determination of over 40 elements in microliter sample volumes with detection limits typically 100 to 1000 times better than those of Flame AA systems.

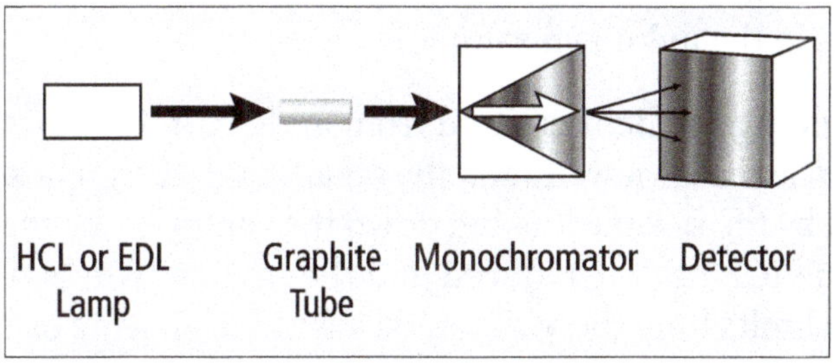

Simplified drawing of a Graphite Furnace AA system.

Elements of the Periodic Table:

1 H Hydrogen 1.00794 — Atomic number; Element Symbol; Element Name; Atomic Mass

1 H Hydrogen 1.00794																	2 He Helium 4.003
3 Li Lithium 6.941	4 Be Beryllium 9.012182											5 B Boron 10.811	6 C Carbon 12.0107	7 N Nitrogen 14.00674	8 O Oxygen 15.9994	9 F Fluorine 18.9984032	10 Ne Neon 20.1797
11 Na Sodium 22.989770	12 Mg Magnesium 24.3050											13 Al Aluminum 26.981538	14 Si Silicon 28.0855	15 P Phosphorus 30.973761	16 S Sulfur 32.066	17 Cl Chlorine 35.4527	18 Ar Argon 39.948
19 K Potassium 39.0983	20 Ca Calcium 40.078	21 Sc Scandium 44.955910	22 Ti Titanium 47.867	23 V Vanadium 50.9415	24 Cr Chromium 51.9961	25 Mn Manganese 54.938049	26 Fe Iron 55.845	27 Co Cobalt 58.933200	28 Ni Nickel 58.6934	29 Cu Copper 63.546	30 Zn Zinc 65.39	31 Ga Gallium 69.723	32 Ge Germanium 72.61	33 As Arsenic 74.92160	34 Se Selenium 78.96	35 Br Bromine 79.904	36 Kr Krypton 83.80
37 Rb Rubidium 85.4678	38 Sr Strontium 87.62	39 Y Yttrium 88.90585	40 Zr Zirconium 91.224	41 Nb Niobium 92.90638	42 Mo Molybdeum 95.94	43 Tc Technetium (98)	44 Ru Ruthenium 101.07	45 Rh Rhodium 102.90550	46 Pd Palladium 106.42	47 Ag Silver 107.8682	48 Cd Cadmium 112.411	49 In Indium 114.818	50 Sn Tin 118.710	51 Sb Antimony 121.760	52 Te Tellurium 127.60	53 I Iodine 126.90447	54 Xe Xenon 131.29
55 Cs Cesium 132.90545	56 Ba Barium 137.327	57 La Lanthanum 138.9055	72 Hf Hafnium 178.49	73 Ta Tantalum 180.9479	74 W Tungsten 183.84	75 Re Rhenium 186.207	76 Os Osmium 190.23	77 Ir Iridium 192.217	78 Pt Platinum 195.078	79 Au Gold 196.96655	80 Hg Mercury 200.59	81 Tl Thallium 204.3833	82 Pb Lead 207.2	83 Bi Bismuth 208.98038	84 Po Polonium (209)	85 At Astatine (210)	86 Rn Radon (222)
87 Fr Francium (223)	88 Ra Radium (226)	89 Ac Actinium (227)	104 Rf Rutherfordium (261)	105 Db Dubnium (262)	106 Sg Seaborgium (263)	107 Bh Bohrium (262)	108 Hs Hassium (265)	109 Mt Meitnerium (266)	110 Ds Darmstadtium (269)	111 Rg Roentgenium (272)							
				58 Ce Cerium 140.116	59 Pr Praseodymium 140.90765	60 Nd Neodymium 144.24	61 Pm Promethium (145)	62 Sm Samarium 150.36	63 Eu Europium 151.964	64 Gd Gadolinium 157.25	65 Tb Terbium 158.92534	66 Dy Dysprosium 162.50	67 Ho Holmium 164.93032	68 Er Erbium 167.26	69 Tm Thulium 168.93421	70 Yb Ytterbium 173.04	71 Lu Lutetium 174.967
				90 Th Thorium 232.0381	91 Pa Protactinium 231.03588	92 U Uranium 238.0289	93 Np Neptunium (237)	94 Pu Plutonium (244)	95 Am Americium (243)	96 Cm Curium (247)	97 Bk Berkelium (247)	98 Cf Californium (251)	99 Es Einsteinium (252)	100 Fm Fermium (257)	101 Md Mendelevium (258)	102 No Nobelium (259)	103 Lr Lawrencium (262)

Inductively Coupled Plasma Optical Emission Spectroscopy

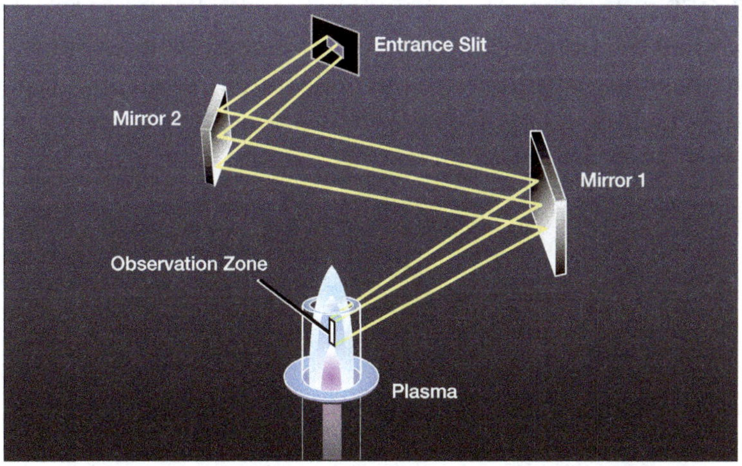

Radially viewed plasma with a vertical slit image in the plasma.

ICP is an argon plasma maintained by the interaction of an RF field and ionized argon gas. The plasma can reach temperatures as high as 10,000 °K, allowing the complete atomization of the elements in a sample and minimizing potential chemical interferences.

Inductively Coupled Plasma Optical Emission Spectroscopy (ICP-OES) is the measurement of the light emitted by the elements in a sample introduced into an ICP source. The measured emission intensities are then compared to the intensities of standards of known concentration to obtain the elemental concentrations in the unknown sample.

There are two ways of viewing the light emitted from an ICP. In the classical ICP-OES configuration, the light across the plasma is viewed radially, resulting in the highest upper linear ranges. By viewing the light emitted by the sample looking down the center of the torch or axially, the continuum background from the ICP itself is reduced and the sample path is maximized. Axial viewing provides better detection limits than those obtained via radial viewing by as much as a factor of 10. The most effective systems allow the plasma to be viewed in either orientation in a single analysis, providing the best detection capabilities and widest working ranges.

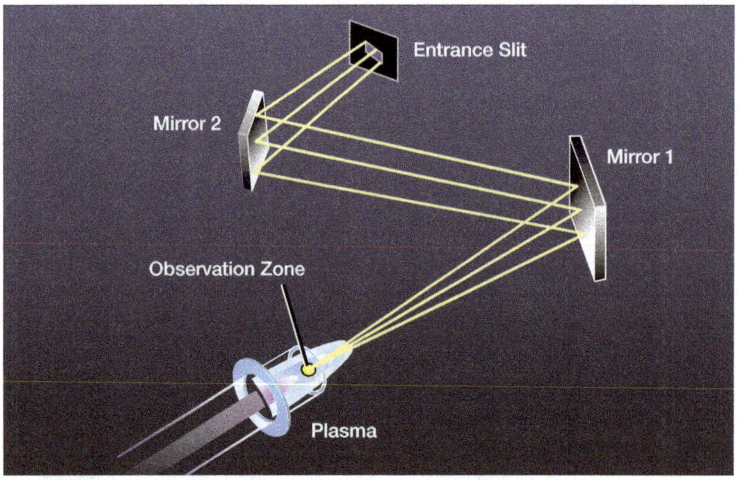

Axially viewed plasma with an axial slit image in the plasma.

The optical system used for ICP-OES consists of a spectrometer that is used to separate the individual wavelengths of light and focus the desired wavelengths onto the detector. Older, "direct reader" types of ICP-OES systems used a series of photomultiplier tubes to determine pre-selected wavelengths. This limited the number of elements that could be determined as the wavelengths were generally fixed once the instrument was manufactured. Sequential-type systems can select any wavelength and focus it on a single detector. However, this is done one element at a time, which can lead to longer analysis times.

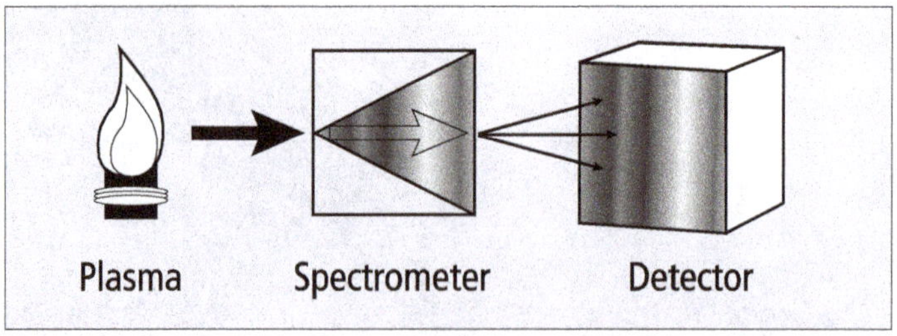

Simplified drawing of a basic ICP system.

In today's modern ICP-OES systems, solid-state detectors based on charge-coupled devices (CCD) are used, providing very flexible systems and eliminating the need for large numbers of single photomultiplier detectors.

Inductively Coupled Plasma Mass Spectrometry

With Inductively Coupled Plasma Mass Spectrometry (ICP-MS), the argon ICP generates singly charged ions from the elemental species within a sample that are directed into a mass spectrometer and separated according to their mass-to-charge ratio. Ions of the selected mass-to-charge ratio are then directed to a detector that determines the number of ions present. Typically, a quadrupole mass spectrometer is used for its ease of-use, robustness and speed. Due to the similarity of the sample-introduction and data-handling techniques, using an ICP-MS is very much like using an ICP-OES system.

ICP-MS combines the multi-element capabilities of ICP techniques with exceptional detection limits equivalent to or below those of GFAA. It is also one of the few analytical techniques that allow the quantification of elemental isotopic concentrations and ratios, as well as precise speciation capabilities when used in conjunction with HPLC or GC interfaces. This feature enables the analytical chemist to determine the exact form of a species present – not just the total concentration.

However, due to the fact that the sample components are actually introduced into the instrument, there are some limitations as to how much sample matrix can be introduced into the ICP-MS. In addition, there are also increased maintenance requirements as compared to ICP-OES systems. Generally, ICP-MS systems require that the total dissolved solids content of a sample be below 0.2% for routine operation and maximum stability. There are several items, such as the interface cones and ion lens, located between the ICP torch and the mass spectrometer, that need to be cleaned on a periodic basis to maintain acceptable instrument performance.

Recent developments have led to new technologies to increase the robustness and stability of ICP-MS. Orthogonal ion lens systems increase the ability of the ICP-MS to handle higher total dissolved solids content and dramatically improve longterm stability for high matrix solutions. Interference control has been made even easier by using universal cell technologies that include both collision (using Kinetic Energy Discrimination KED) and Dynamic Reaction Cell (DRC) in a single instrument allowing the analyst to choose the best technique for their samples.

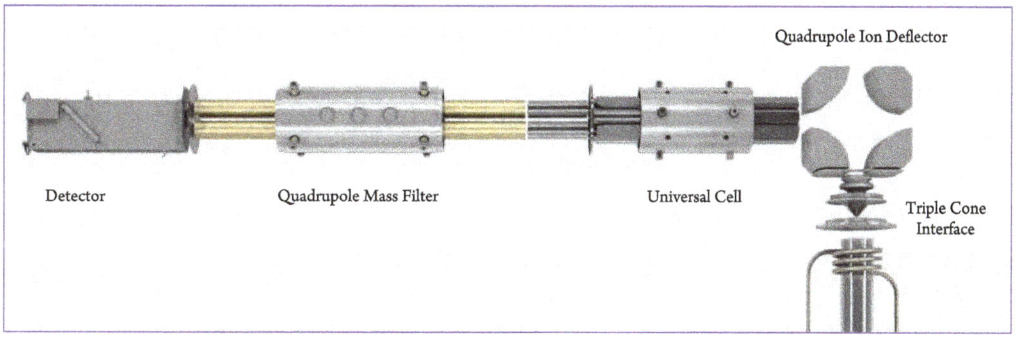

Simplified drawing of ICP-MS system with Universal Cell Technology (UCT).

Atomic Spectra

When atoms are excited they emit light of certain wavelengths which correspond to different colors. The emitted light can be observed as a series of colored lines with dark spaces in between; this series of colored lines is called a line or atomic spectra. Each element produces a unique set of spectral lines. Since no two elements emit the same spectral lines, elements can be identified by their line spectrum.

Electromagnetic Radiation and the Wave Particle Duality

Energy can travel through a vacuum or matter as electromagnetic radiation. Electromagnetic radiation is a transverse wave with magnetic and electric components that oscillate perpendicular to each other. The electromagnetic spectrum is the range of all possible wavelengths and frequencies of electromagnetic radiation including visible light.

According to the wave particle duality concept, although electromagnetic radiation is often considered to be a wave, it also behaves like a particle. In 1900, while studying black body radiation, Max Planck discovered that energy was limited to certain values and was not continuous as assumed in classical physics. This means that when energy increases, it does so by tiny jumps called quanta (quantum in the singular). In other words, a quantum of energy is to the total energy of a system as an atom is to the total mass of a system. In 1905, Albert Einstein proposed that energy was bundled into packets, which became known as photons. The discovery of photons explained why energy increased in small jumps. If energy was bundled into tiny packets, each additional packet would contribute a tiny amount of energy causing the total amount of energy to jump by a tiny amount, rather than increase smoothly as assumed in classical physics.

- λ is the wavelength of light.
- v is the frequency of light.
- n is the quantum number of an energy state.
- E is the energy of that state.

Table: Important Constants.

Constant	Meaning	Value
c	Speed of light	$2.99792458 \times 10^{8}\ ms^{-1}$
h	Planck's constant	6.62607×10^{-34} Js
eV	Electron volt	1.60218×10^{-19} J
R_H	Rydberg constant for H	2.179×10^{-18} J

Wavelength, or the distance from one peak to the other of a wave, is most often measured in meters, but can be measured using other SI units of length where practical. The number of waves that pass per second is the frequency of the wave. The SI unit for frequency is the Hertz (abbreviated Hz). 1 Hz is equal to $1s^{-1}$. The speed of light is constant. In a vacuum the speed of light is

2.99792458 × 10^8 ms^{-1}. The relationship between wavelength (λ), frequency (v), and the speed of light (c) is:

$$v = \frac{c}{\lambda}$$

The energy of electromagnetic radiation of a particular frequency is measured in Joules and is given by the equation:

$$E = hv$$

with *h* as Planck's constant (6.62606876 × 10^{-34} Js).

The electron volt is another unit of energy that is commonly used. The electron volt (eV) is defined as the kinetic energy gained by an electron when it is accelerated by a potential electrical difference of 1 volt. It is equal to 1.60218 × 10^{-19} J.

Spectroscope

A spectrum is a range of frequencies or wavelengths. By the process of refraction, a prism can split white light into it's component wavelengths. However this method is rather crude, so a spectroscope is used to analyze the light passing through the prism more accurately. The diagram below shows a simple prism spectroscope. The smaller the difference between distinguishable wavelengths, the higher the resolution of the spectroscope. The observer sees the radiation passing through the slit as a spectral line. To obtain accurate measurements of the radiation, and electronic device often takes the place of the observer, the device is then called a spectrophotometer. In more modern Spectro-photometers, a diffraction grating is used instead of a prism to disperse the light.

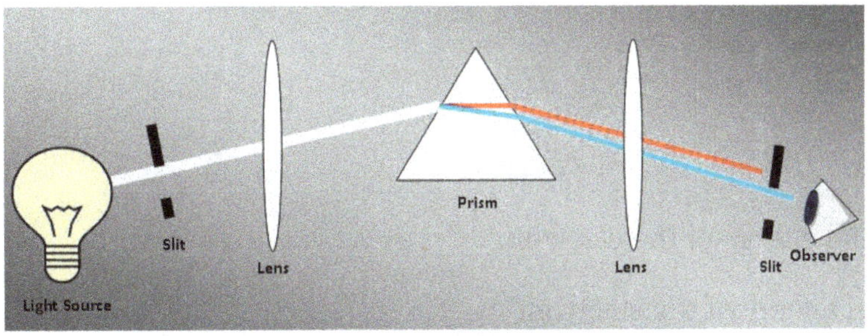

Simple Prism Spectroscope.

How Atoms React when Excited by Light

Electrons can only exist in certain areas around the nucleus called shells. Each shell corresponds to a specific energy level which is designated by a quantum number n. Since electrons cannot exist between energy levels, the quantum number n is always an integer value (n = 1, 2, 3, 4...). The electron with the lowest energy level (n = 1) is the closest to the nucleus. An electron occupying its lowest energy level is said to be in the ground state. The energy of an electron in a certain energy level can be found by the equation:

$$E_n = \frac{-R_H}{n_2}$$

Where R_H is a constant equal to 2.179×10^{-18} J and n is equal to the energy level of the electron.

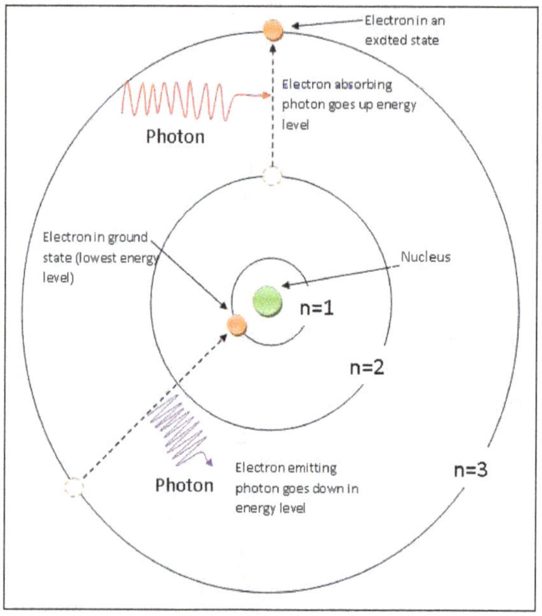

When light is shone on an atom, its electrons absorb photons which cause them to gain energy and jump to higher energy levels. The higher the energy of the photon absorbed, the higher the energy level the electron jumps to. Similarly, an electron can go down energy levels by emitting a photon. The simplified version of this principal is illustrated in the figure below is based on the Bohr model of the Hydrogen atom. The energy of the photon emitted or gained by an electron can be calculated from this formula:

$$E_{photon} = R_H \left(\frac{1}{n_i^2} - \frac{1}{n_f^2} \right)$$

with,

- E_i is the initial energy of the electron,
- E_f is the final energy of the electron.

Since an electron can only exist at certain energy levels, they can only emit photons of certain frequencies. These specific frequencies of light are then observed as spectral lines. Similarly, a photon has to be of the exact wavelength the electron needs to jump energy levels in order to be absorbed, explaining the dark bands of an absorption spectra.

Emission Lines

When an electron falls from one energy level in an atom to a lower energy level, it emits a photon of a particular wavelength and energy. When many electrons emit the same wavelength of photons it will result in a spike in the spectrum at this particular wavelength, resulting in the banding pattern seen in atomic emission spectra. The graphic is a simplified picture of a spectrograph, in this case being used to photograph the spectral lines of Hydrogen.

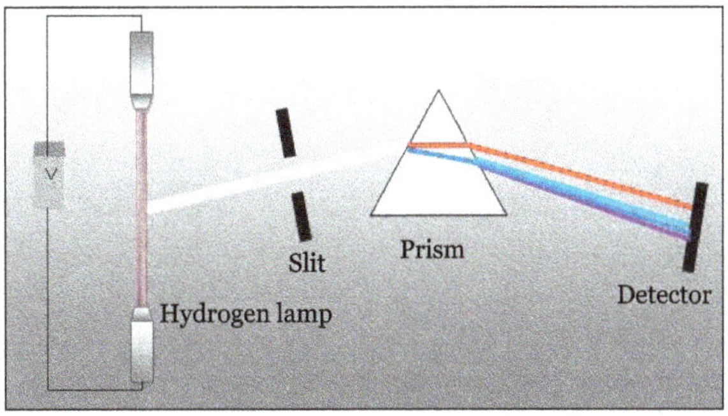

Simplified Spectrograph.

In this spectrograph, the Hydrogen atoms inside the lamp are being excited by an electric current. The light from the lamp then passes through a prism, which diffracts it into its different frequencies. Since the frequencies of light correspond to certain energy levels (n) it is therefore possible to predict the frequencies of the spectral lines of Hydrogen using an equation discovered by Johann Balmer.

$$v = 3.2881 \times 10^{15} s^{-1} \left(\frac{1}{2^2} - \frac{1}{n^2} \right)$$

Where *n* must be a number greater than 2. This is because Balmer's formula only applies to visible light and some longer wavelengths of ultraviolet.

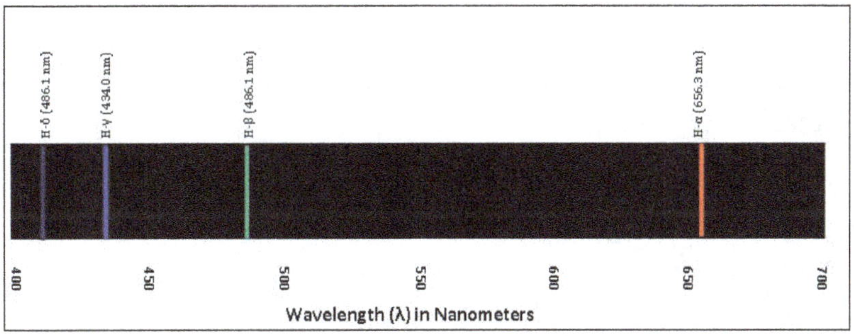

Balmer series for Hydrogen Atom.

The frequencies in this region of Hydrogen's atomic spectra are called the Balmer series. The Balmer series for Hydrogen is pictured above. There are several other series in the Hydrogen atom which correspond to different parts of the electromagnetic spectrum. The Lyman series, for example, extends into the ultraviolet, and therefore can be used to calculate the energy of to n = 1.

Absorption Lines

When an electron jumps from a low energy level to a higher level, the electron will absorb a photon of a particular wavelength. This will show up as a drop in the number of photons of this wavelength and as a black band in this part of the spectrum. The figure below illustrates a mechanism to detect an absorption spectrum. A white light is shone through a sample. The atoms in the sample absorb some of the light, exciting their electrons. Since the electrons only absorb light of certain frequencies, the absorption spectrum will show up as a series of black bands on an otherwise continuous spectrum.

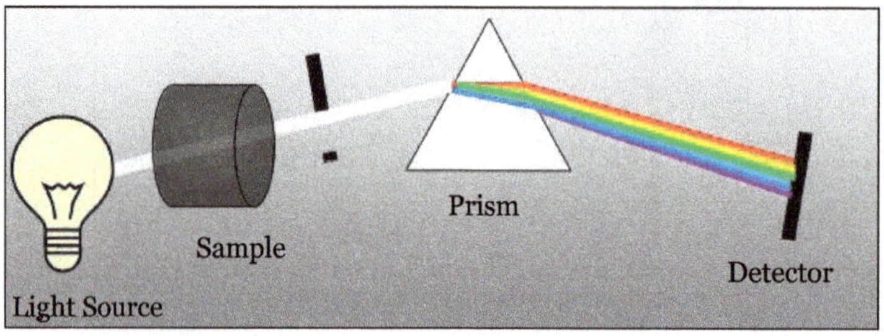

Applications of Atomic Spectral Analysis

Atomic spectroscopy has many useful applications. Since the emission spectrum is different for every element, it acts as an atomic fingerprint by which elements can be identified. Some elements were discovered by the analysis of their atomic spectrum. Helium, for example, was discovered while scientists were analyzing the absorption spectrum of the sun. Emission spectra is especially useful to astronomers who use emission and absorption spectra to determine the make up of far away stars and other celestial bodies.

Alkali Spectra

The absorption spectra of alkali vapors (Such as lithium, sodium) appear quite similar in many respects to the absorption spectrum of H atom. They are only displaced to a considerable extent, toward longer wavelengths. These spectra also consist of a series of lines with regularly decreasing separation and decreasing intensity.

It cannot, however, be represented by a formula completely analogous to the Bohr formula. On the other hand, since the lines converge to a limit, we must be able to represent them as differences between two terms. Rydberg formula,

$$v = T_{PS} - \frac{R}{(m+p)^2}, \text{ where } m = 2,3$$

p is a constant, known as Quantum Defect. T_{PS} is known as series limit. This series is known as Principal series.

Other series, in addition to this, may be observed for the alkalis. They are diffuse, sharp and Bergmann series.

- Sharp Series: $v = T_{SS} - \frac{R}{(m+s)^2} \quad m = 2,3$
- Diffuse Series: $v = T_{SS} - \frac{R}{(m+d)^2} \quad m = 3,4$
- Bergmann Series: $vT_{BS} - \frac{R}{(m+f)^2} \quad m = 4,5$
- Selection rules: $\Delta n = 0,1,2,3, \ \Delta \ell = \pm 1$

As an example, sodium energy levels and transitions are given in figure below.

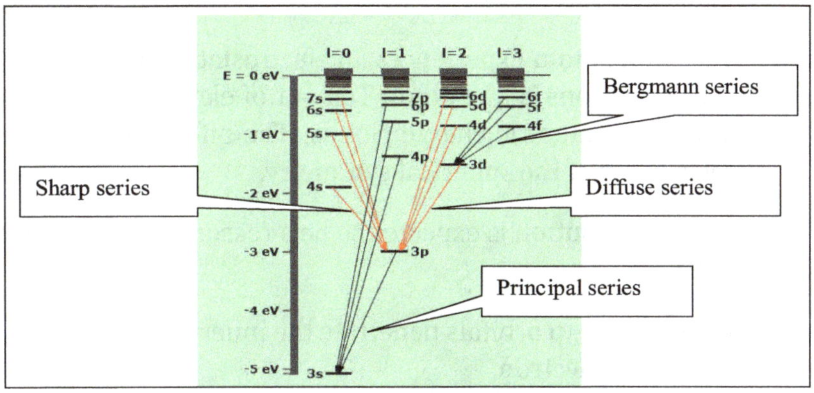

Sodium atom energy levels and transitions.

As a specific example, we consider the alkali metals such as lithium, sodium and potassium, which come from group I of the periodic table. They have one valence electron outside filled inner shells. They are therefore approximately one-electron systems, and can be understood by introducing a phenomenological number called the quantum defect to describe the energies.

Let us consider the sodium atom. The optical spectra are determined by excitations of the outermost 3s electron. The energy of each $(n; l)$ term of the valence electron is given by:

$$E_{n,l} = -\frac{R}{(n-\delta(l))^2}$$

where n ≥ 3, $\delta(l)$ is the quantum defect.

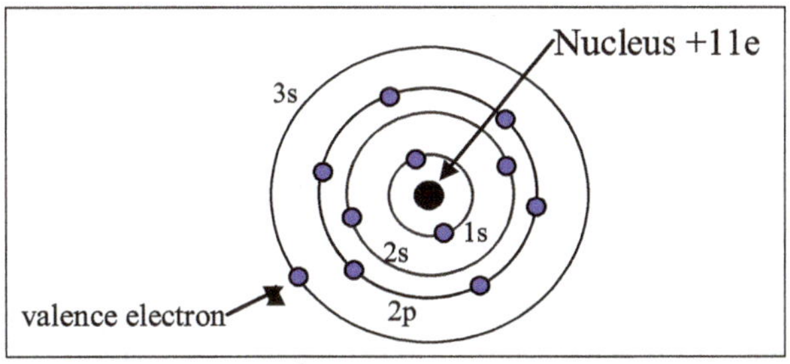

Sodium Atom Z = 11.

The quantum defect $\delta(l)$ was introduced empirically to account for the optical spectra. In principle it should depend on both n and l, but it was found experimentally to depend mainly on l as given in the following table.

Values of quantum defect for sodium				
l	n=3	n = 4	n = 5	n = 6
0	1.373	1.357	1.352	1.349
1	0.883	0.867	0.862	0.859
2	0.010	0.011	0.013	0.011
3	-	0.000	-0.001	-0.008

The dependence of the quantum defect on l can be understood with reference to the figure where the radial probability densities for the 3s and 3p orbitals of a hydrogenic atom with Z = 1 are plotted with respect to normalized radial distance.

An individual electron in sodium atom experiences an electrostatic potential due to the Coulomb repulsion from all the other electrons in the atom. Ten out of eleven electrons are in closed subshells, which have spherically-symmetric charge clouds. The off-radial forces from electrons in these closed shells cancel because of the spherical symmetry.

Hydrogen radial probability distribution is expected to be a reasonable approximation for the single valence electron of sodium.

- We see that both the 3s and 3p orbitals penetrate the inner shells, and that this penetration is much greater for the 3s electron.
- The electron will therefore see a larger effective nuclear charge for part of its orbit, and this will have the effect of reducing the energies.
- The energy reduction is largest for the 3s electron due to its larger core penetration.

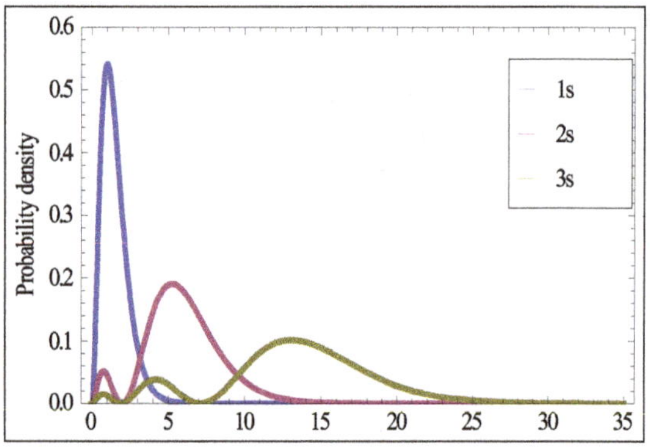

Probability Density.

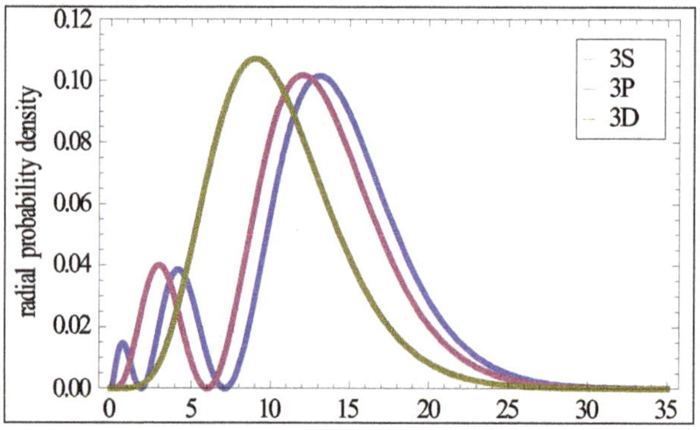

Radial Probability densities.

The effect of this penetration results in the shift of energy levels. A comparison with the hydrogen energy level is shown in figure.

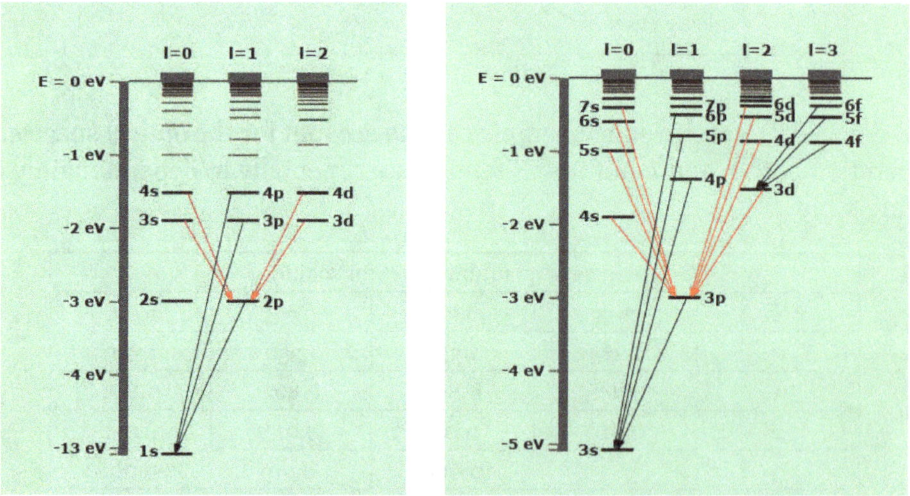

Hydrogen energy levels and sodium energy levels.

We can say that,

- Energy levels with different ℓ have different energies. In other words λ degeneracy removed.
- From the hydrogen atom energy levels, it cannot be described, because energy depends on "n" only.

Classical Explanation

Penetrating and Non-Penetrating Orbits as shown in figure.

- Non-Penetrating orbits: The first is the case when the outer electron has a non-penetrating orbit, as in the figure. If it is accepted that the mean symmetry of the cloud formed by (Z −1) electrons is similar, the electron experiences the electrostatic potential of the nuclear charge of Ze and of the spherical distribution of charge (Z −1). The discussion presented for the hydrogen atom remains valid.

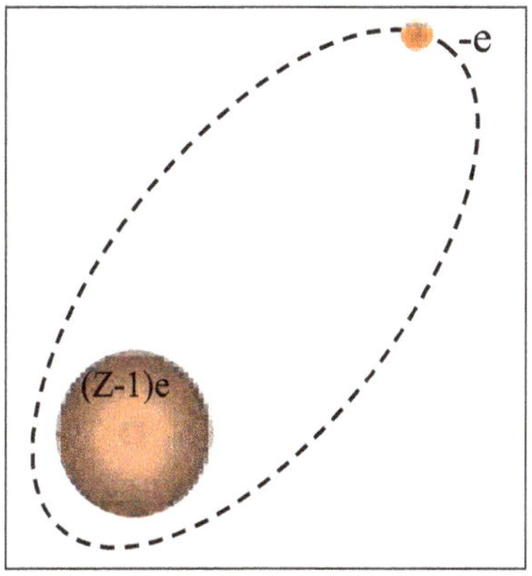

- Penetrating orbit: On the other hand, if the orbit of the outer electron penetrates inside the core of the atom, the problem is much more complex, simple solution by Somerfield, is this,

$$V_{ext} = \frac{1}{4\pi\varepsilon_0}\frac{e}{r} \qquad "r" \text{ Large}$$

$$V_{in} = \frac{1}{4\pi\varepsilon_0}\frac{Ze}{r} + \text{Constant} \qquad "r" \text{ Small}$$

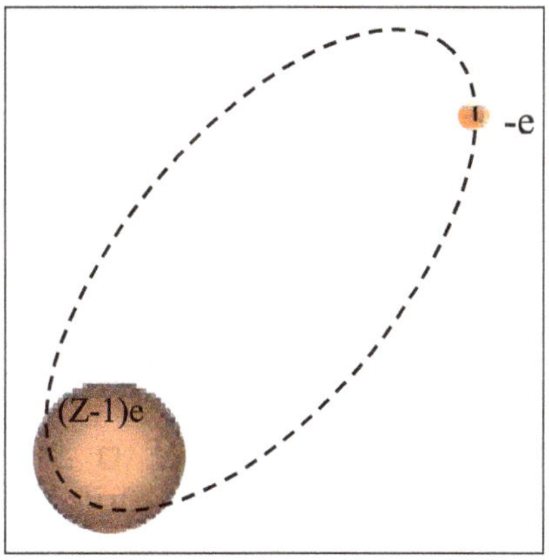

Quantum Mechanical Calculation

Form of the potential energy, $V(r) = -\dfrac{1}{4\pi\varepsilon_0}\dfrac{e^2}{r}\left(1+\dfrac{b}{r}\right)$

This form represents the potential energy requirement at large distance,

$$V(r) = -\frac{1}{4\pi\varepsilon_0}\frac{e^2}{r}$$

and at small distance, $V(r) = -\dfrac{1}{4\pi\varepsilon_0}\dfrac{Ze^2}{r}$

This potential with respect to radial distance is shown in figure below.

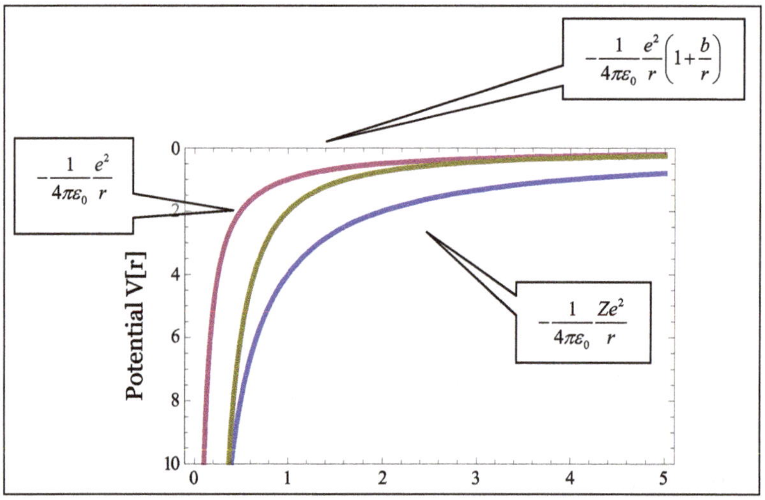

Radial distance.

Since this is radial dependence and we need to solve only radial equation of the Schrödinger equation of hydrogen atom problem,

$$V(r) = -\frac{1}{4\pi\varepsilon_0}\frac{e^2}{r}\left(1+\frac{b}{r}\right)$$
$$= \frac{c}{r}\left(1+\frac{b}{r}\right)$$

where,

$$c = -\frac{1}{4\pi\varepsilon_0}e^2$$

Now, the Hamiltonian for one electron atom,

$$H = -\frac{\hbar}{2\mu}\nabla^2 + V(r)$$

$$= -\frac{\hbar^2}{2\mu}\nabla^2 + \frac{c}{r}\left(1+\frac{b}{r}\right)$$

$$\Rightarrow -\frac{\hbar^2}{2\mu}\nabla^2\psi + \frac{c}{r}\left(1+\frac{b}{r}\right)\psi = E\psi$$

$$\Rightarrow \nabla^2\psi + \frac{2\mu}{\hbar^2}\left[E - \frac{c}{r}\left(1+\frac{b}{r}\right)\right]\psi = 0$$

The radial equation:

$$\frac{d^2\chi}{dr^2} + \left[\frac{2\mu}{\hbar^2}E - \frac{2\mu}{\hbar^2}\frac{c}{r}\left(1+\frac{b}{r}\right) - \frac{\ell(\ell+1)}{r^2}\right]R(r) = 0$$

$$\Rightarrow \frac{d^2\chi}{d^2} - \left[-\frac{2\mu}{r\hbar^2}E + \frac{2\mu}{\hbar^2}\frac{c}{r} + \frac{2\mu}{\hbar^2}\frac{cb}{r^2} + \frac{\ell(\ell+1)}{r^2}\right]R(r) = 0$$

Taking $A = -\frac{2\mu E}{\hbar^2}$; $B = -\frac{2\mu c}{\hbar^2}$ and substituting in equation above,

$$\frac{d^2\chi}{dr^2} - \left[A - \frac{B}{r} + \frac{-Bb + \ell(\ell+1)}{r^2}\right]\chi = 0$$

Let,

$$\ell^*(\ell^*+1) = -Bb + \ell(\ell+1)$$

So,

$$\frac{d^2\chi}{dr^2} - \left[A - \frac{B}{r} + \frac{-Bb + \ell^*(\ell^*+1)}{r^2}\right]\chi = 0$$

Same radial equation as in hydrogen atom, solution with n^* where $n^* = \ell^* + p + 1$,

$$E = -\frac{R}{(n^*)^2} = -\frac{R}{(\ell^*+p+1)^2}$$

$$n = \ell + p + 1$$

so,

$$n^* = \ell^* + p + 1 = \ell^* - \ell + n$$

$$n^* = n - (\ell - \ell^*) = n - \Delta\ell$$

Now,

$$\ell^*(\ell^*+1) = \ell(\ell+1) - Bb$$
$$\Rightarrow (\ell^2 - (\ell^*)^2) + (\ell - \ell^*) = Bb$$
$$\Rightarrow (\ell - \ell^*)(\ell + \ell^* + 1) = Bb$$
$$\Rightarrow \Delta\ell = \frac{Bb}{\ell + \ell^* + 1} = \frac{Bb}{2\ell + 1}$$
$$= \frac{B/_{2b}}{\ell + 1/2} = \frac{b}{a_1}\frac{1}{\ell + 1/2}$$
$$E_{n,\ell} = -\frac{Rhc}{\left(n - \frac{b}{a_1}\frac{1}{\ell + 1/2}\right)^2}$$

Note that:

This energy expression is dependent on both n and ℓ.

Maximum $\ell \rightarrow$ small correction.

Small $\ell \rightarrow$ correction term is large.

$$E_{n,\ell} = -\frac{Rhc}{\left\{n - \frac{Bb}{2\ell + 1}\right\}^2}$$

Now we know,

$$E = hv$$
$$v = \frac{c}{\lambda}, \qquad \bar{v} = \frac{1}{\lambda}$$
$$E = \frac{hc}{\lambda} \Rightarrow \frac{1}{\lambda}\left(cm^{-1}\right) = \frac{E}{hc}$$
$$E_{n,\ell} = -\frac{R}{\left\{n - \frac{Bb}{2\ell + 1}\right\}^2}c$$

where R = 109,728.7 cm^{-1}.

Conversion

	Ergs/molecule	Cal/molecule	Electron volts
1 cm^{-1}	1.9858×10^{-16}	2.8575	1.23954×10^{-4}

Take an example:

Lithium: Ionization potential – 43,486 cm^{-1} or 5.39 cm eV,

$$43486 = \frac{109728.7}{(2 - Bb)^2}$$

$$\Rightarrow (2 - Bb)^2 = 2.5233$$

$$\Rightarrow 2 - Bb = 1.588$$

$$\Rightarrow Bb = 0.41$$

$$E_{2,1} = \frac{109728.7}{\left(2 - \frac{0.41}{2}\right)^2} = 31603.8\,cm^{-1}$$

$$E_{3,0} = \frac{109728.7}{(3 - 0.41)^2} = 16357.6\,cm^{-1}$$

$$E_{3,1} = \frac{109728.7}{\left(3 - \frac{0.41}{3}\right)^2} = 13383.7\,cm^{-1}$$

$$E_{4,0} = \frac{109728.7}{(4 - 0.41)^2} = 8513.9\,cm^{-1}$$

$$E_{4,1} = \frac{109728.7}{\left(4 - \frac{0.41}{3}\right)^2} = 7351.8\,cm^{-1}$$

$$E_{4,2} = \frac{109728.7}{\left(4 - \frac{0.41}{5}\right)^2} = 7148.1\,cm^{-1}$$

$$E_{4,3} = \frac{109728.7}{\left(4 - \frac{0.41}{7}\right)^2} = 7063.4\,cm^{-1}$$

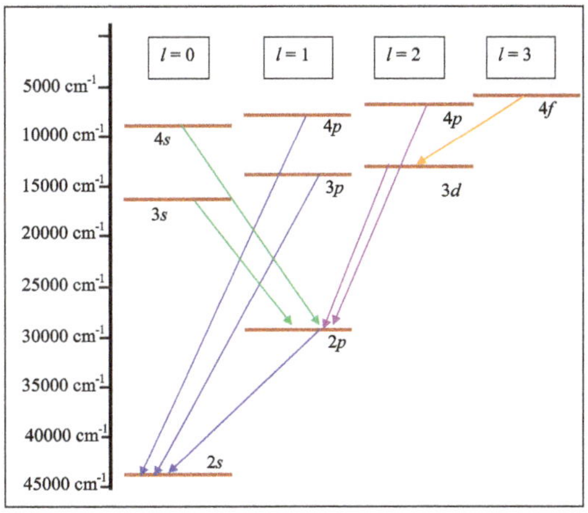

Doublet Structure of Alkali Spectra

Careful examination of the spectra of alkali metals shows that each member of some of the series are closed doublets. For example, sodium yellow line, corresponding to 3p → 3s transition, is a close doublet with separation of 6 Å while potassium (K) has a doublet separation of 34 Å and so on. Further investigations show that only the S-terms are singlet, while all the other terms P, D, F etc. are doublets. Such doublet structure in energy is observed for all the atoms possessing a single valence electron i.e., in the outer most shell. Usually the doublet spacing is small compared to the term difference (for Na the main D line is cantered at 5893 Å; D_2 = 5890 Å and D_1 = 5896 Å) and hence it is called fine structure. To explain this feature Uhlenback and Goudsmith first proposed the hypothesis of electron spin, which was later on obtained as a natural consequence of Dirac's relativistic theory.

Spin is essentially a quantum phenomenon. The spin of the electron is found to be $\frac{1}{2}\hbar$ and $S^2 = s(s+1)\hbar^2$ where $s = \frac{1}{2}$, the quantum number for spin – (Li, Na, K, Rb, Cs, Fr).

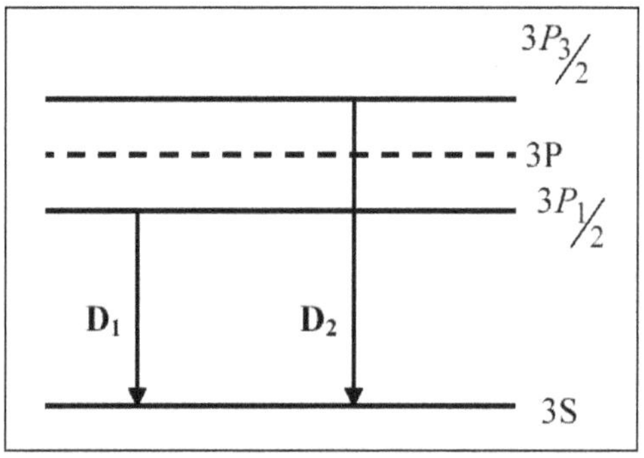

Explanation of doublet structure of alkali atom: The Hamiltonian for the lone electron of an alkali atom relative to the atomic core is given by,

$$H_0 = \frac{p^2}{2\mu} + v(r)$$

where,

- $\vec{P}$ = momentum of the lone electron.
- $V(\vec{r})$ = Potential.
- $\vec{r}$ = Distance of the lone electron from the centre of the atomic core, i.e., nucleus.

Now considering the spin orbit interaction the total Hamiltonian is given by,

$$H = \frac{p^2}{2\mu} + V(r) + H^{s-o}$$

where H^{s-o} = spin orbit interaction.

An electron with orbital angular momentum $\vec{l}$ and spin $\vec{s}$ will behave a total angular momentum,

$$\vec{j} = \vec{l} + \vec{s}$$

which gives the quantum numbers for $\vec{j}$ as $j=(l+s),....,(l-s)$. Since for a single electron $s=\frac{1}{2}$ or $-\frac{1}{2}, \therefore j=l+\frac{1}{2}$ and $l-\frac{1}{2}$, for $l=0; j=\frac{1}{2}$ only.

The coupling of spin with orbital angular momentum gives rise to the fine structure splitting of spectral lines and it gives a full account of the doublet structure of alkali spectra. This interaction is called spin orbit interaction.

The existence of electron spin and the doublet structure comes as a natural consequence of the relativistic theory but classical electrodynamics gives a non-realistic description.

Consider an electron of charge –e moving with velocity $\vec{v}$ at a distance $\vec{r}$ from the nucleus of charge Ze (or an entire atomic configuration of effective charge Ze).

Viewed from electrons rest frame the entire atomic configuration (consisting of the nucleus and the rest of the electronic charge cloud be taken as rigid core) is moving with a velocity of – V and their effective field also moves with the same velocity. Associated with the moving electric field arising from relativistic transformation equation. To first order in $\frac{v}{c}$ the magnetic field is $\vec{B}=\frac{1}{c}\vec{E}\times\vec{v}$ and, $\vec{E}=-\vec{\nabla}\phi; \phi=$ scalar potential,

$$\therefore B=-\frac{1}{c}\vec{\nabla}\Phi\times\vec{v}$$

$$=+\frac{1}{ec}\vec{\nabla}V\times\vec{v}$$

where,

$$V=-e\phi=\text{potential.}$$

Now for a spherical symmetric potential,

$$\nabla=\frac{dV}{dr}\frac{r}{|r|}$$

$$\therefore \quad =-\frac{1}{ec}\frac{1}{r}\frac{dV}{dr}\vec{r}\times\vec{v}$$

$$=\frac{1}{mec}\frac{1}{r}\frac{dV}{dr}\vec{r}\times\vec{p} \quad [\vec{p}=m\vec{v}]$$

$$=\frac{1}{mec}\frac{1}{r}\frac{dV}{dr}l \quad [l=\vec{r}\times\vec{p}=\text{angular momentum.}]$$

If the coordinate system is fixed with the atom the effective magnetic field is to be multiplied by a factor $\frac{1}{2}$. This is known as Thomas correction obtained from a spinning top model of the electron.

$$\therefore \mathrm{B} = \frac{1}{2mec} \cdot \frac{1}{r}\frac{dV}{dr}\vec{l}$$

The interaction energy of the electronic magnetic moment with the magnetic field due to the motion of charge particle is:

$$H^{s-o} = -\vec{\mu}_s . \vec{\mathrm{B}}$$

$$= \frac{e}{mc} \cdot \frac{1}{2mec} \cdot \frac{1}{r}\frac{dV}{dr}\vec{l} . \vec{s} \left[\mu_s = -\frac{e}{mc}\vec{s} \right]$$

This equation holds in the rest frame of both electron and the nucleus as to first order in $\frac{v}{c}$, the energy is same. There is no term proportional to l^2 as in the rest frame of the electron; there is no orbital magnetic moment.

Now we can write using quantum mechanical operator for $\vec{l}$ and $\vec{s}$ $(as\, \hbar\vec{l}$ and $\hbar\vec{s})$.

$$\therefore H^{s-o} = \frac{\hbar^2}{2m^2c^2} \cdot \frac{1}{r}\frac{dV}{dr}\vec{l} . \vec{s}$$

$$= \xi(r)\vec{l} . \vec{s}$$

where $\vec{l}$ and $\vec{s}$ are dimensionless vector operators,

$$\xi(r) = \frac{\hbar^2}{2m^2c^2} \cdot \frac{1}{r}\frac{dv}{dr}$$

The total Hamiltonian of the alkali atom considering spin orbit interaction is given by,

$$H = \frac{p^2}{2\mu} + V(r) + \xi(r)(l \cdot \vec{s})$$

For a pure Coulombic potential,

$$V(r) = -\frac{Ze^2}{r}$$

$$\therefore \frac{1}{r}\frac{dV}{dr} = \frac{Ze^2}{r^3}$$

$$\therefore \xi(r) = \frac{\hbar^2 Ze^2}{2m^2c^2} \cdot \frac{1}{r^3}$$

Now the spin orbit term can be considered as perturbation term. The contribution of the spin orbit term can be calculated by considering its expectation value,

$$\therefore \Delta E = \left\langle \psi \middle| H^{s-o} \middle| \psi \right\rangle$$

where ψ is the ground state wave function of the effective one particle system.

In an atom the total angular momentum $\vec{j}$ is always conserved even if individual $\vec{l}$ and $\vec{s}$ may not be conserved. Hence we can work in a representation or system where H^{s-o} is diagonal. Also we choose $\left|nl_j m_j\right\rangle$ representation, where individual $\vec{l}$ and $\vec{s}$ are combined to form the conserved quantity $\vec{j}$.

Thus,

$$\begin{aligned}\Delta E &= \left\langle \psi_{nljm_j} \left| H^{s-o} \right| \psi_{nljm_j} \right\rangle \\ &= \left\langle \psi_{nl}(r) \middle| \xi(r) \middle| \psi_{nl}(r) \right\rangle \left\langle \psi_{lsjm_j} \left| \vec{l}\cdot\vec{s} \right| \varphi_{lsjm_j} \right\rangle \\ &= \varsigma_{nl} \left\langle \vec{l}\cdot\vec{s} \right\rangle.\end{aligned}$$

Where,

$$\begin{aligned}\varsigma_{nl} &= \left\langle \psi_{nl}(r) \middle| \xi(r) \middle| \psi_{nl}(r) \right\rangle \\ &= \int_0^\infty R_{nl}^2(r)\xi(r) r^2 dr \\ &= \frac{Ze^2\hbar^2}{2m^2c^2}\int_0^\infty R_{nl}^2(r)\frac{1}{r^3}r^2 dr \\ &= \frac{Ze^2\hbar^2}{2m^2c^2}\left\langle \frac{1}{r^3} \right\rangle\end{aligned}$$

we have,

$$\left\langle \frac{1}{r^3} \right\rangle \frac{Z^3}{n^3 l(l+1)\left(l+\frac{1}{2}\right)a_0^3}, \qquad a_0 \frac{\hbar^2}{me^2}$$

$\therefore$ We can write,

$$\varsigma_{nl} = R_y a^2 \frac{Z^4}{n^3 l(l+1)\left(l+\frac{1}{2}\right)}$$

where,

$$R_y = \frac{me^4}{2\hbar^2} = \text{Rydberg constant,}$$

and

$$a = \frac{e^2}{\hbar c} = \text{fine structure constant.}$$

To calculate the angular term $\langle \vec{l} \cdot \vec{s} \rangle$ we proceed as follows:

$$\vec{l} \cdot \vec{s} = \frac{1}{2}\left[\vec{j}^2 - \vec{l}^2 - \vec{s}^2\right]$$

$$\therefore \left\langle \psi_{lsjm_j} \middle| \vec{l} \cdot \vec{s} \middle| \psi_{lsjm_j} \right\rangle = \left\langle \vec{l} \cdot \vec{s} \right\rangle$$

$$= \frac{1}{2}\langle | j^2 - l^2 - s^2 | \rangle$$

$$= \frac{1}{2}\left[j(j+1) - l(l+1) - s(s+1) \right]$$

$$= \frac{1}{2}\left[j(j+1) - l(l+1) - \frac{3}{4} \right], \quad \because s = \frac{1}{2}$$

Here for a given $l, j = l + \frac{1}{2}$ and $l - \frac{1}{2}$.

Thus each level split into two sublevels with $j = l + \frac{1}{2}$ and $l - \frac{1}{2}$.

For,

$$\hat{j} = l + \frac{1}{2}$$

$$\left\langle \vec{l} \cdot \vec{s} \right\rangle = \frac{1}{2}\left[\left(l + \frac{1}{2}\right)\left(l + \frac{3}{2}\right) - l(l+1) - \frac{3}{4} \right]$$

$$= \frac{1}{2}\left[l^2 + \frac{3l}{2} + \frac{1}{2}l + \frac{3}{4} - l^2 - l - \frac{3}{4} \right]$$

$$= \frac{l}{2}$$

For,

$$\hat{j} = l - \frac{1}{2}$$

$$\left\langle \vec{l} \cdot \vec{s} \right\rangle = \frac{1}{2}\left[\left(l + \frac{1}{2}\right)\left(l + \frac{3}{2}\right) - l(l+1) - \frac{3}{4} \right]$$

$$= \frac{1}{2}\left[l^2 - \frac{1}{4} - l^2 - l - \frac{3}{4} \right]$$

$$= -\frac{1}{2}(l+1)$$

Thus the energy levels are given by,

$$E_1\left(\hat{j} = l + \frac{1}{2}\right) = E_0 + \zeta_{nl}\frac{1}{2}$$

$$E_2\left(\hat{j}=l-\frac{1}{2}\right)=E_0-\zeta_{nl}\frac{(l+1)}{2}$$

The separation is given by,

$$\Delta E=E_1-E_2=\zeta_{nl}\left(l+\frac{1}{2}\right)$$

Thus the splitting of energy levels i.e., separation between the two levels for each value of l is $\frac{1}{2}(2l+1)\zeta_{nl}$. This is the so called fine structure of spectral lines for spin orbit interaction of energy levels.

Now we may draw the energy level diagram for the non-relativistic theory then the correction to the energy levels due to relativistic variation of mass and at last correction to the energy level due to the presence of spin orbit interaction.

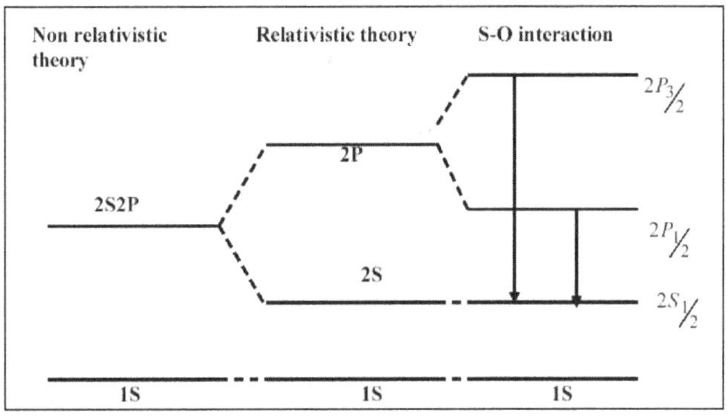

Now for $l=0, j=\pm\frac{1}{2}$ but j can never be –ve. Hence $j=+\frac{1}{2}$. Therefore $1s$ level doesn't split up for (S-O) interaction.

For,

$$l=1, j=\frac{1}{2},\frac{3}{2}$$

$$l=2, j=\frac{5}{2},\frac{3}{2}$$

$$l=3, j=\frac{7}{2},\frac{5}{2} \text{ and so on.}$$

Thus except l = o (which has no S-O interaction) the other energy levels split up into two levels (doubles) as shown below due to S-O interaction.

The mechanism responsible to the doublet splitting is S-O interaction. Each energy level has got a spin multiplicity $\left(2s+\frac{1}{2}\right)=2\cdot\frac{1}{2}+1=1+1=2.$

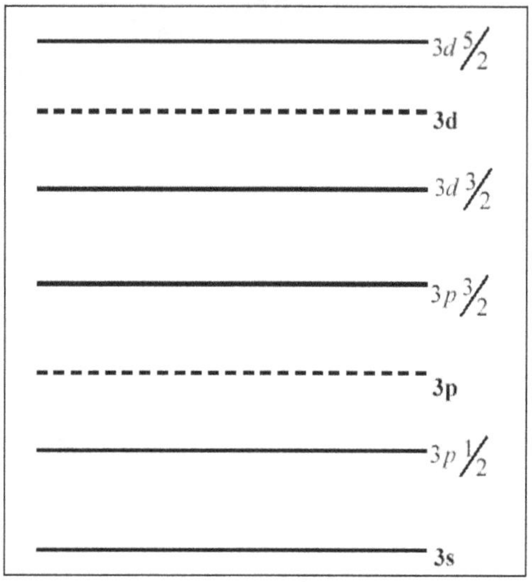

In spectroscopic notation, each energy level is represented as ${}_{n}^{2s+1}L_{j}$,

- n = Principal quantum number.
- L = orbital angular momentum quantum number.
- J = Total angular momentum quantum number $= l + s$.
- 2 +1 = spin multiplicity.
- l = 0, 1, ...n – 1.
- m_l = -1.......... +1.
- j = l + s.
- j = (1 + s), (1 + s – 1), ... (1 – s).

Fine Structure

Fine structure is the splitting of the main spectral lines of an atom into two or more components, each representing a slightly different wavelength. Fine structure is produced when an atom emits light in making the transition from one energy state to another. The split lines, which are called the fine structure of the main lines, arise from the interaction of the orbital motion of an electron with the quantum mechanical "spin" of that electron. An electron can be thought of as an electrically charged spinning top, and hence it behaves as a tiny bar magnet. The spinning electron interacts with the magnetic field produced by the electron's rotation about the atomic nucleus to generate the fine structure.

The amount of splitting is characterized by a dimensionless constant called the fine-structure constant. This constant is given by the equation $\alpha = ke^2/hc$, where k is Coulomb's constant, e is the charge of the electron, h is Planck's constant, and c is the speed of light. The value of the constant α is $7.29735254 \times 10^{-3}$, which is nearly equal to 1/137.

In the atoms of alkali metals such as sodium and potassium, there are two components of fine structure (called doublets), while in atoms of alkaline earths there are three components (triplets). This arises because the atoms of alkali metals have only one electron outside a closed core, or shell, of electrons, while the atoms of alkaline earths have two such electrons. Doublet separation for corresponding lines increases with atomic number; thus, with lithium (atomic number 3), a doublet may not be resolved by an ordinary spectroscope, whereas with rubidium (atomic number 37), a doublet may be widely separated.

Fine Structure of Hydrogen

The fine structure of hydrogen is the spectrum of the atom. After the partial simplifications we have,

$$H = \underbrace{\frac{\hat{p}^2}{2m} + \mathrm{V}}_{H^{(0)}} - \underbrace{\frac{\hat{p}^4}{8m^3c^2}}_{\delta H_{rel.}} + \underbrace{\frac{e^2}{2m^2c^2}\frac{\mathrm{S.L}}{r^3}}_{\delta H_{\text{spin-orbit}}} + \underbrace{\frac{\pi e^2\hbar^2}{2m^2c^2}\delta(r)}_{\delta H_{\text{Darwin}}}.$$

Darwin Correction

Let us now evaluate the Darwin correction. Since this interaction has a delta function at the origin, the first order correction to the energy vanishes unless the wavefunction is non-zero at the origin. This can only happen for nS states. There is no need to use the apparatus of degenerate perturbation theory. Indeed, for fixed n there are two orthogonal ℓ = 0 states, one with electron spin up and one with electron spin down. While these states are degenerate, the Darwin perturbation commutes with spin and is therefore diagonal in the two-dimensional subspace. There is no need to include the spin in the calculation and we have,

$$E^{(1)}_{n00,\text{Darwin}} = \langle \psi_{n00} | \delta H_{\text{Darwin}} | \psi_{n00} \rangle = \frac{\pi e^2\hbar^2}{2m^2c^2} |\psi_{n00}(0)|^2 .$$

The radial equation can be used to determine the value of the nS wave functions at the origin. You will find that,

$$|\psi_{n00}(0)|^2 = \frac{1}{\pi n^3 a_0^3}.$$

As a result,

$$E^{(1)}_{n00,\text{Darwin}} = \frac{e^2\hbar^2}{2m^2c^2a_0^3n^3} = \alpha^4(mc^2)\frac{1}{2n^3}.$$

This completes the evaluation of the Darwin correction.

The Darwin term in the Hamiltonian arises from the elimination of one of the two two component spinors in the Dirac equation. As we will show now such a correction would arise from a nonlocal correction to the potential energy term. It is as if the electron had grown from point-like to a ball with radius of order its Compton wavelength $\frac{\hbar}{m_e c}$. The potential energy due to the field of the proton must then be calculated by integrating the varying electric potential over the charge distribution of the electron. While a simple estimate of this nonlocal potential energy does reproduce the Darwin correction rather closely, one must not reach the conclusion that the electron is no longer a point particle. Still the fact remains that in a relativistic treatment of an electron, its Compton wavelength is relevant and is physically the shortest distance an electron can be localized.

The potential energy $V(r)$ of the electron, as a point particle, is the product of the electron charge $(-e)$ times the electric potential $\Phi(r)$ created by the proton:

$$V(r) = (-e)\Phi(r) = (-e)\frac{e}{r}.$$

Let us call $\tilde{V}(r)$ the potential energy when the electron is a charge distribution centered at a point r with $|r| = r$.

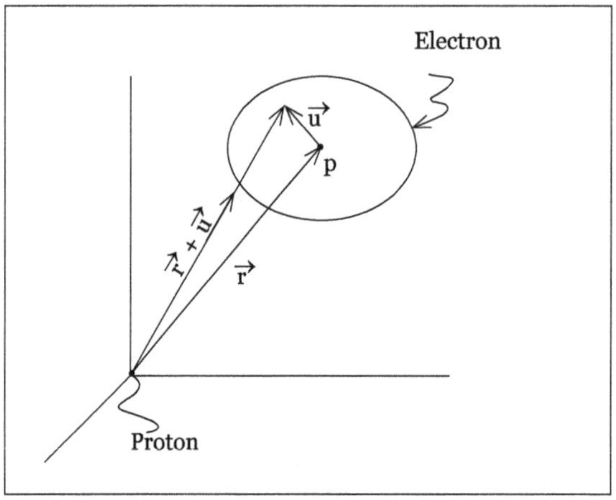

A Darwin type correction to the energy arises if the electron charge is smeared over a region of size comparable to its Compton wavelength. Here the center of the spherically symmetric electron cloud is at P and the proton is at the origin. The vector u is radial relative to the center of the electron.

This energy is obtained by integration over the electron distribution. Using the vector u to define position relative to the center P of the electron, and letting $\rho(u)$ denote the position dependent charge density, we have:

$$\tilde{V}(r) = \int_{\text{electron}} d^3\text{u}\, \rho(\text{u})\Phi(r + \text{u}),$$

where, as shown in the figure, $r + u$ is the position of the integration point, measured relative to the proton at the origin. It is convenient to write the charge density in terms of a normalized function ρ_o:

$$\rho(\text{u}) = -e\rho_0(\text{u}) \rightarrow \int_{\text{electron}} d^3\text{u}\, \rho_0(\text{u}) = 1,$$

which guarantees that the integral of ρ over the electron is indeed $(-e)$. Recalling that $-e\Phi(\text{r}+\text{u}) = \text{V}(\text{r}+\text{u})$, we now rewrite equation $\tilde{V}(r) = \int_{\text{electron}} d^3\text{u}\, \rho(\text{u})\Phi(r+\text{u})$ as:

$$\tilde{V}(r) = \int_{\text{electron}} d^3\text{u}\, \rho_0(\text{u})V(r + \text{u}).$$

This equation has a clear interpretation: the potential energy is obtained as a weighted integral of potential due to the proton over the extended electron. If the electron charge would be perfectly localized, $\rho_0(\mathrm{u}) = \delta(\mathrm{u})$ and $\tilde{V}(r)$ would just be equal to *V*(*r*). We will assume that the distribution of charge is spherically symmetric, so that:

$$\rho_0(\mathrm{u}) = \rho_0(\mathrm{u}).$$

To evaluate equation $\tilde{V}(\mathrm{r}) = \int d^3\,\mathrm{u}\,\rho_0(\mathrm{u})\left(V(\mathrm{r}) + \sum_i \partial_i V\Big|_{\mathrm{r}} u_i + \frac{1}{2}\sum_{i,j} \partial_i\partial_j V\Big|_{\mathrm{r}} u_i u_j + \ldots \right)$,

we first do a Taylor expansion of the potential that enters the integral about the point u = 0:

$$V(r+\mathrm{u}) = V(r) + \sum_i \partial_i V\Big|_{\mathrm{r}} u_i + \frac{1}{2}\sum_{i,j} \partial_i\partial_j V\Big|_{\mathrm{r}} u_i u_j + \ldots$$

All derivatives here are evaluated at the center of the electron. Plugging back into the integral and dropping the subscript 'electron' we have,

$$\tilde{V}(\mathrm{r}) = \int d^3\,\mathrm{u}\,\rho_0(\mathrm{u})\left(V(\mathrm{r}) + \sum_i \partial_i V\Big|_{\mathrm{r}} u_i + \frac{1}{2}\sum_{i,j} \partial_i\partial_j V\Big|_{\mathrm{r}} u_i u_j + \ldots \right)$$

All r dependent functions can be taken out of the integrals. Recalling that the integral of ρ_o over volume is one, we get:

$$\tilde{V}(\mathrm{r}) = V(\mathrm{r}) + \sum_i \partial_i V\Big|_{\mathrm{r}} \int d^3\,\mathrm{u}\,\rho_0(\mathrm{u})\rho_i + \frac{1}{2}\sum_{i,j} \partial_i\partial_j V\Big|_{\mathrm{r}} \int d^3\,\mathrm{u}\,\rho_0(\mathrm{u}) u_i u_j + \ldots$$

Due to spherical symmetry the first integral vanishes and the second takes the form,

$$\int d^3\,\mathrm{u}\,\rho_0(\mathrm{u}) u_i u_j = \frac{1}{3}\delta_{ij}\int d^3\,\mathrm{u}\,\rho_0(\mathrm{u}) u^2.$$

Indeed the integral must vanish for $i \neq j$ and must take equal values for $i = j = 1,2,3$. Since $u^2 = u_1^2 + u_2^2 + u_3^2$, the result follows.

Using this we get,

$$\tilde{V}(\mathrm{r}) = V(\mathrm{r}) + \frac{1}{2}\sum_i \partial_i\partial_j V\Big|_{\mathrm{r}} \frac{1}{3}\int d^3\,\mathrm{u}\, f(\mathrm{u})\rho^2 + \ldots$$

$$= V(\mathrm{r}) + \frac{1}{6}\nabla^2\nabla \int d^3\,\mathrm{u}\,\rho_0(\mathrm{u}) u^2 + \ldots$$

The second term represents the correction δV to the potential energy:

$$\delta V = \frac{1}{6}\nabla^2\nabla \int d^3\,\mathrm{u}\,\rho_0(\mathrm{u}) u^2.$$

To get an estimate, let us assume that the charge is distributed uniformly over a sphere of radius u_o. This means that $\rho_o(u)$ is a constant for $u < u_o$.

$$\rho_0(\mathrm{u}) = \frac{3}{4\pi u_0^3}\begin{cases}1, & u < u_0,\\ 0, & u > u_0.\end{cases}$$

The integral one must evaluate then gives,

$$\int d^3\,\mathrm{u}\,\rho_0(\mathrm{u})u^2 = \int_0^{u_0}\frac{4\pi u^2 du\,u^2}{\frac{4\pi}{3}u_0^3} = \frac{3}{u_0^3}\int_0^{u_0}u^4 du = \frac{3}{5}u_0^2.$$

Therefore,

$$\delta V = \frac{1}{10}u_0^2\nabla^2\nabla.$$

If we choose the radius u_0 of the charge distribution to be the Compton wavelength $\frac{\hbar}{mc}$ of the electron we get,

$$\delta V = \frac{\hbar^2}{10m^2c^2}\nabla^2 V.$$

Comparing with equation,

$$H = \underbrace{\frac{\hat{p}^2}{2m} + \mathrm{V}}_{H^{(0)}} - \underbrace{\frac{\hat{p}^4}{8m^3c^2}}_{\delta H_{rel.}} + \underbrace{\frac{1}{2m^2c^2}\frac{1}{r}\frac{dV}{dr}\mathrm{S.L}}_{\delta H_{\text{spin-orbit}}} + \underbrace{\frac{\hbar^2}{8m^2c^2}\nabla^2\,\mathrm{V}}_{\delta H_{\text{Darwin}}},$$

we see that, up to a small correction $\left(\frac{1}{8}\text{ as opposed to }\frac{1}{10}\right)$, this is the Darwin energy shift. The agreement is surprisingly good for what is, admittedly, a heuristic argument.

Relativistic Correction

We now turn to the relativistic correction. The energy shifts of the hydrogen states can be analyzed among the degenerate states with principal quantum number n. We write tentatively for the corrections:

$$E^{(1)}_{n,\ell m_\ell m_s;\mathrm{rel}} = -\frac{1}{8m^3c^2}\left\langle \psi_{n\ell m_\ell m_s}\right|\mathrm{P}^2\mathrm{P}^2\left|\psi_{n\ell m_\ell}\right\rangle.$$

We can use this formula because the uncoupled basis of states at fixed n is good: the perturbation is diagonal in this basis. This is clear because the perturbing operator p^2p^2 commutes with L^2, with L_z, and with S_z. The first operator guarantees that that the matrix for the perturbation is diagonal in ℓ, the second guarantees that the perturbation is diagonal in mℓ, and the third guarantees, rather trivially, that the perturbation is diagonal in m_s.

To evaluate the matrix element we use the Hermiticity of p^2 to move one of the factors into the bracket,

$$E^{(1)}_{n,\ell m_\ell m_s;\mathrm{rel}} = -\frac{1}{8m^3c^2}\left\langle \mathrm{P}^2\psi_{n\ell m} \middle| \mathrm{P}^2\psi_{n\ell m}\right\rangle,$$

where in the right-hand side we evaluated the trivial expectation value for the spin degrees of freedom. To simplify the evaluation we use the Schrödinger equation, which tells us that,

$$\left(\frac{\mathrm{p}^2}{2m}+V\right)\psi_{n\ell m} = E_n^{(0)}\psi_{n\ell m} \rightarrow \mathrm{p}^2\psi_{n\ell m} = 2m(E_n^{(0)}-V)\psi_{n\ell m}.$$

Using this both for the bra and the ket:

$$E^{(1)}_{n,\ell m_\ell m_s;\mathrm{rel}} = -\frac{1}{2mc^2}\left\langle (E_n^{(0)}-V)\psi_{n\ell m} \middle| E_n^{(0)}-V)\psi n\ell m\right\rangle.$$

The operator $E_n^{(0)}-V$ is also Hermitian and can be moved from the bra to the ket, giving,

$$E^{(1)}_{n,\ell m_\ell m_s;\mathrm{rel}} = -\frac{1}{2mc^2}\left\langle \psi_{n\ell m}\middle|((E_n^{(0)})^2-2VE_n^{(0)}+V^2)\middle|\psi_{n\ell m}\right\rangle$$

$$= -\frac{1}{2mc^2}\left[(E_n^{(0)})^2-2E_n\langle V\rangle_{n\ell m}+\langle V^2\rangle_{n\ell m}\right].$$

The problem has been reduced to the computation of the expectation value of $V(\mathrm{I})$ and $V^2(r)$ in the $\psi_{n\ell m}$ state. The expectation value of $V(r)$ is obtained from the virial theorem that states that $\langle V\rangle = 2E_n^{(0)}$. For $V^2(\mathrm{r})$ we have,

$$\langle V^2\rangle = e^4\left\langle\frac{1}{r^2}\right\rangle = e^4\frac{1}{a_0^2n^3(\ell+\frac{1}{2})} = \left(\frac{e^2}{2a_0}\frac{1}{n^2}\right)^2\frac{4n}{\ell+\frac{1}{2}} = (E_n^{(0)})^2\cdot\frac{4n}{\ell+\frac{1}{2}}.$$

Back into equation $E^{(1)}_{n,\ell m_\ell m_s;\mathrm{rel}} = -\frac{1}{2mc^2}\left\langle \psi_{n\ell m}\middle|((E_n^{(0)})^2-2VE_n^{(0)}+V^2)\middle|\psi_{n\ell m}\right\rangle$

$$= -\frac{1}{2mc^2}\left[(E_n^{(0)})^2-2E_n\langle V\rangle_{n\ell m}+\langle V^2\rangle_{n\ell m}\right]$$

$$E^{(1)}_{n,\ell m_\ell m_s;\mathrm{rel}} = -\frac{(E_n^{(0)})^2}{2mc^2}\left[\frac{4n}{\ell+\frac{1}{2}}-3\right] = -\frac{1}{8}\alpha^4\frac{(mc^2)}{n^4}\left[\frac{4n}{\ell+\frac{1}{2}}-3\right].$$

The complete degeneracy of ℓ multiplets for a given n has been broken. That degeneracy of $H^{(0)}$ was explained by the conserved Runge-Lenz vector. It is clear that the relativistic correction has broken that symmetry.

We have computed the above correction using the uncoupled basis,

$$E^{(1)}_{n,\ell m_\ell m_s;\mathrm{rel}} = \langle n\ell m m_s | \delta H_{\mathrm{rel}} | n\ell m m_s \rangle = f(n,\ell).$$

Here we added the extra equality to emphasize that the matrix elements depend only on n and ℓ. We have already seen that in the full degenerate subspace with principal quantum number n the matrix for δH_{rel} is diagonal in the uncoupled basis. But now we see that in each degenerate subspace of fixed n and ℓ, δH_{rel} is in fact a multiple of the identity matrix, since the matrix elements are independent of m and m_s (the L_z and S_z) eigenvalues. A matrix equal to a multiple of the identity is invariant under any orthonormal change of basis. For any $\ell \otimes \frac{1}{2}$ multiplet, the resulting *j* multiplets provide an alternative orthonormal basis. The invariance of a matrix proportional to the identity implies that,

$$E^{(1)}_{n,\ell m_\ell m_s,\mathrm{rel}} = \langle n\ell j m_j | \delta H_{\mathrm{rel}} | n\ell j m_j \rangle = f(n,\ell).$$

with the same function *f*(*n*, ℓ) as in equation $E^{(1)}_{n,\ell m_\ell m_s;\mathrm{rel}} = \langle n\ell m m_s | \delta H_{\mathrm{rel}} | n\ell m m_s \rangle = f(n,\ell)$, and the perturbation is diagonal in this coupled basis too. This is clear anyway because the perturbation commutes with L^2 , J^2 and J^z and and any two degenerate states in the coupled basis differ either in ℓ, j or j^z.

The preservation of the matrix elements can also be argued more explicitly. Indeed, any state in the coupled basis is a superposition of orthonormal uncoupled basis states with constant coefficients c_i:

$$|n\ell j m_j\rangle = \sum_i c_i |n\ell m_\ell^i m_s^i\rangle, \text{ with } \sum_i |c_i|^2 = 1,$$

because the state on the left-hand side must also have unit norm. Therefore, using the diagonal nature of the matrix elements in the uncoupled basis we get, as claimed,

$$\begin{aligned}\langle n\ell j m_j | \delta H_{\mathrm{rel}} | n\ell j m_j \rangle &= \sum_{i,k} c_i^* c_k \langle n\ell m_\ell^i m_s^i | \delta H_{\mathrm{rel}} | n\ell m_\ell^k m_s^k \rangle \\ &= \sum_i |c_i|^2 \langle n\ell m_\ell^i m_s^i | \delta H_{\mathrm{rel}} | n\ell m_\ell^i m_s^i \rangle \\ &= \sum_i |c_i|^2 f(n,\ell) = f(n,\ell).\end{aligned}$$

Spin Orbit Coupling

The spin-orbit contribution to the Hamiltonian is,

$$\delta H \text{ spin-orbit} = \frac{e^2}{2m^2c^2}\frac{1}{r^3}\mathbf{S}\cdot\mathbf{L}.$$

$\delta H_{\text{spin-orbit}}$ commutes with L^2 because L^2 commutes with any $\hat{L}_i$ and any $\hat{S}_i$. Moreover, $\delta H_{\text{spin-orbit}}$ commutes with J^2 and with j_z since, in fact, $[\hat{J}_i, S\cdot L] = 0$ for any i; $S\cdot L$ is a scalar operator for J. It follows that $\delta H_{\text{spin-orbit}}$ is diagonal in the level n degenerate subspace in the coupled basis $|n\ell j m_j\rangle$. In fact, as we will see, the matrix elements are m-independent. This is a nontrivial consequence of $\delta H_{\text{spin-orbit}}$ being a scalar under J. To compute the matrix elements we recall that J = S + L,

$$E^{(1)}_{n\ell jm_j;\text{spin-orbit}} = \frac{e^2}{2m^2c^2}\left\langle n\ell jm_j\left|\frac{1}{r^3}S\cdot L\right|n\ell jm_j\right\rangle$$
$$= \frac{e^2}{2m^2c^2}\frac{\hbar^2}{2}[j(j+1)-\ell(\ell+1)-\frac{3}{4}]\left\langle n\ell jm_j\left|\frac{1}{r^3}\right|n\ell jm_j\right\rangle.$$

We need the expectation value of $1/r^3$ in these states. It is known that,

$$\left\langle n\ell m_\ell\left|\frac{1}{r^3}\right|n\ell m_\ell\right\rangle = \frac{1}{n^3a_0^3\ell(\ell+\frac{1}{2})(\ell+1)}.$$

Because of the m_ℓ independence of this expectation value (and its obvious m_s independence) the operator $1/r^3$ is a multiple of the identity matrix in each $\ell\otimes\frac{1}{2}$ multiplet. It follows that it is the same multiple of the identity in the coupled basis description. Therefore,

$$\left\langle njm_j\left|\frac{1}{r^3}\right|n\ell jm_j\right\rangle = \frac{1}{n^3a_0^3\ell(\ell+\frac{1}{2})(\ell+1)}.$$

Using this in,

$$E^{(1)}_{n\ell jm_j;\text{spin-orbit}} = \frac{e^2\hbar^2}{4m^2c^2}\frac{[j(j+1)-\ell(\ell+1)-\frac{3}{4}]}{n^3a_0^3\ell(\ell+\frac{1}{2})(\ell+1)}.$$

Working out the constants in terms of $E_n^{(0)}$ and rest energies we get,

$$E^{(1)}_{n\ell jm_j;\text{spin-orbit}} = \frac{(E_n^{(0)})^2}{mc^2}\frac{n[j(j+1)-\ell(\ell+1)]-\frac{3}{4}}{\ell(\ell+\frac{1}{2})(\ell+1)},\quad \ell\neq 0.$$

Since L vanishes identically acting on any $\ell = 0$ state, it is physically reasonable, as we will do, to assume that the spin-orbit correction vanishes for $\ell = 0$ states. On the other hand the limit of the above formula as $\ell \to 0$, while somewhat ambiguous, is nonzero. We set $j = \ell + \frac{1}{2}$ (the other possibility $j = \ell - \frac{1}{2}$ does not apply for $\ell = 0$) and then take the limit as $\ell \to 0$. Indeed,

$$E^{(1)}_{n\ell j m_j;\text{spin-orbit}}\Bigg|_{j=\ell+\frac{1}{2}} = \frac{(E_n^{(0)})^2}{mc^2}\frac{n\left[(\ell+\frac{1}{2})(\ell+\frac{3}{2})-\ell(\ell+1)-\frac{3}{4}\right]}{\ell(\ell+\frac{1}{2})(\ell+1)}$$

$$= \frac{(E_n^{(0)})^2}{mc^2}\frac{n}{(\ell+\frac{1}{2})(\ell+1)},$$

and now taking the limit:

$$\lim_{\ell\to 0} E^{(1)}_{n\ell j m_j;\text{spin-orbit}}\Bigg|_{j=\ell+\frac{1}{2}} = \frac{(E_n^{(0)})^2}{mc^2}(2n) = \alpha^4 mc^2 \frac{1}{2n^3}.$$

We see that this limit is in fact identical to the Darwin shift of the nS states.

Combining Results

For $\ell \neq 0$ states we can add the energy shifts from spin-orbit and from the relativistic correction, both of them expressed as expectation values in the coupled basis. The result, therefore will give the shifts of the coupled states. Collecting our results and we have,

$$\langle n\ell j m_j | \delta H_{\text{rel}} + \delta H_{\text{spin-orbit}} | n\ell j m_j \rangle$$

$$= \frac{(E_n^{(0)})^2}{2mc^2}\left\{3 - \frac{4n}{(\ell+\frac{1}{2}} + \frac{2n[j(j+1)-\ell(\ell+1)-\frac{3}{4}]}{\ell(\ell+\frac{1}{2})(\ell+1)}\right\}$$

$$= \frac{(E_n^{(0)})}{2mc^2}\left\{3 + 2n\left[\frac{j(j+1)-3\ell(\ell+1)-\frac{3}{4}}{\ell(\ell+\frac{1}{2})(\ell+1)}\right]\right\}$$

These are the fine structure energy shifts for all states in the spectrum of hydrogen. The states in a coupled multiplet are characterized by ℓ, j and m_j and each multiplet as a whole is shifted according to the above formula. The degeneracy within the multiplet is unbroken because the formula has no m_j dependence. This formula, as written, hides some additional degeneracies.

In the above formula there are two cases to consider for any fixed value of j: the multiple can have $\ell = j - \frac{1}{2}$ or the multiplet can have $\ell = j + \frac{1}{2}$. We will now see something rather surprising. In both of these cases the shift is the same, meaning that the shift is in fact ℓ independent! It just depends on j. Call $f(j, \ell)$ the term in brackets above,

$$f(j,\ell) \equiv \frac{j(j+1)-3\ell(\ell+1)-\frac{3}{4}}{\ell(\ell+\frac{1}{2})(\ell+1)}.$$

The evaluation of this expression in both cases gives the same result:

$$f(j,\ell)\bigg|_{\ell=j-\frac{1}{2}} = \frac{j(j+1)-3(j-\frac{1}{2})(j+\frac{1}{2})-\frac{3}{4}}{(j-\frac{1}{2})j)(j+\frac{1}{2})} = \frac{-2j^2+j}{j(j-\frac{1}{2})(j+\frac{1}{2})} = -\frac{2}{(j+\frac{1}{2})},$$

$$f(j,\ell)\bigg|_{\ell=j-\frac{1}{2}} = \frac{j(j+1)-3(j+\frac{1}{2})(j+\frac{3}{2})-\frac{3}{4}}{(j+\frac{1}{2})j)(j+1)(j+\frac{3}{2})} = \frac{2j^2-5j-3}{(j+\frac{1}{2})(j+1)(j+\frac{3}{2})} = -\frac{2}{(j+\frac{1}{2})}.$$

We can therefore replace in $\frac{(E_n^{(0)})^2}{2mc^2}\left\{3-\frac{4n}{(\ell+\frac{1}{2}}+\frac{2n[j(j+1)-\ell(\ell+1)-\frac{3}{4}]}{\ell(\ell+\frac{1}{2})(\ell+1)}\right\} = \frac{(E_n^{(0)})}{2mc^2}\left\{3+2n\left[\frac{j(j+1)-3\ell(\ell+1)-\frac{3}{4}}{\ell(\ell+\frac{1}{2})(\ell+1)}\right]\right\}$ the result of our evaluation, which we label as fine structure (fs) shifts:

$$E^{(1)}_{n\ell j,m_j;\text{fs}} = -\frac{(\text{E}_n^{(0)})^2}{2mc^2}\left[\frac{4n}{j+\frac{1}{2}}-3\right] = -\alpha^4(mc^2)\frac{1}{2n^4}\left[\frac{n}{j+\frac{1}{2}}-\frac{3}{4}\right].$$

More briefly we can write,

$$E^{(1)}_{n\ell j,m_j;\text{fine}} = -\alpha^4 mc^2\cdot S_{n,j},\ \text{with}\ S_{n,j} \equiv \frac{1}{2n^4}\left[\frac{n}{j+\frac{1}{2}}-\frac{3}{4}\right].$$

Let us consider a few remarks:

1. The dependence on j and absence of dependence on ℓ in the energy shifts could be anticipated from the Dirac equation. The rotation generator that commutes with the Dirac Hamiltonian is J = L+S, which simultaneously rotates position, momenta, and spin states. Neither L nor S are separately conserved. With J a symmetry, states are expected to be labelled by energy and j and must be m_j independent.
2. The formula works for nS states. For these $\ell = 0$ states we were supposed to add the relativistic correction and the Darwin correction, since their spin-orbit correction is zero. But we noticed that the limit $\ell \rightarrow 0$ of the spin-orbit correction reproduces the Darwin term. Whether or not this is a meaningful coincidence, it means the sum performed above gives the right answer for $\ell \rightarrow 0$.
3. While a large amount of the degeneracy of $H^{(0)}$ has been broken, for fixed n, multiplets with the same value of j, regardless of ℓ, remain degenerate. The states in each j multiplet do not split.
4. Since $S_{n,j} > 0$ all energy shifts are down. Indeed,

$$\frac{n}{j+\frac{1}{2}} \geq \frac{n}{j_{\max}+\frac{1}{2}} = \frac{n}{\ell_{\max}+\frac{1}{2}+\frac{1}{2}} = \frac{n}{n} = 1 \rightarrow \frac{n}{j+\frac{1}{2}} - \frac{3}{4} \geq \frac{1}{4}.$$

5. For a given fixed n, states with lower values of j get pushed further down. As n increases splittings fall off like n^{-3}.

A table of values of $S_{n,j}$ is given here below.

n	j	$S_{n,j}$
1	$\frac{1}{2}$	$\frac{1}{8}$

2	$\frac{1}{2}$	$\frac{5}{128}$
	$\frac{3}{2}$	$\frac{1}{128}$
3	$\frac{1}{2}$	$\frac{1}{72}$
	$\frac{3}{2}$	$\frac{1}{216}$
	$\frac{5}{2}$	$\frac{1}{648}$

The energy diagram for states up to $n = 3$ is given here,

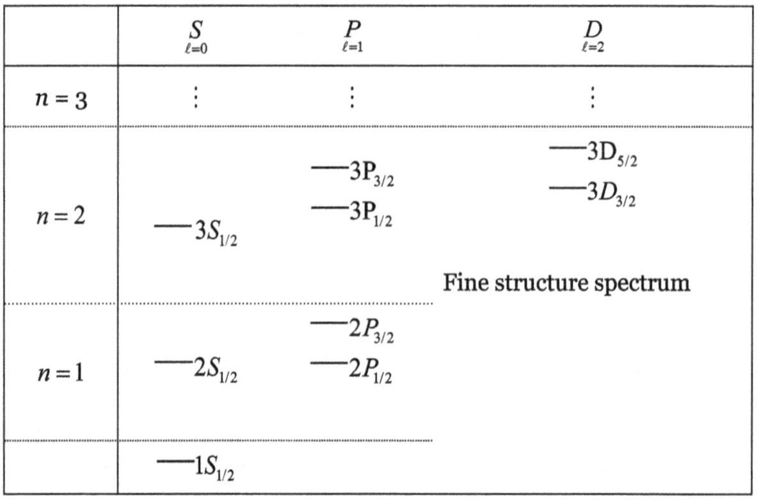

For the record, the total energy of the hydrogen states is the zeroth contribution plus the fine structure contribution. Together they give,

$$E_{n\ell jm_j} = -\frac{e^2}{2a_0}\frac{1}{n^2}\left[1+\frac{\alpha^2}{n^2}\left(\frac{n}{j+\frac{1}{2}}-\frac{3}{4}\right)\right].$$

This is the fine structure of hydrogen. There are, of course, finer corrections. The socalled Lamb shift, for example, breaks the degeneracy between $2S_{1/2}$ and $2P_{1/2}$ and is of order α^5. There is also hyperfine splitting, which arises from the coupling of the magnetic moment of the proton to the magnetic moment of the electron. Such coupling leads to a splitting that is a factor m_e/m_p smaller than fine structure.

Hyperfine Structure

Hyperfine structure (HFS) is the splitting of a spectral line into a number of components. The splitting is caused by nuclear effects and cannot be observed in an ordinary spectroscope without the aid of an optical device called an interferometer. In fine structure, line splitting is the result of energy changes produced by electron spin–orbit coupling (*i.e.*, interaction of forces from orbital and spin motion of electrons); but in hyperfine structure, line splitting is attributed to the fact that in addition to electron spin in an atom, the atomic nucleus itself spins about its own axis. Energy states of the atom will be split into levels corresponding to slightly different energies. Each of these energy levels may be assigned a quantum number, and they are then called quantized levels. Thus, when the atoms of an element radiate energy, transitions are made between these quantized energy levels, giving rise to hyperfine structure.

The spin quantum number is zero for nuclei of even atomic number and even mass number, and therefore no HFS is found in their spectral lines. The spectra of other nuclei do exhibit hyperfine structure. By observing HFS, it is possible to calculate nuclear spin.

A similar effect of line splitting is caused by mass differences (isotopes) of atoms in an element and is called isotope structure, or isotope shift. These spectral lines are sometimes referred to as hyperfine structure but may be observed in an element with spin-zero isotopes (even atomic and mass numbers). Isotope structure is seldom observed without true HFS accompanying it.

Hyperfine Structure of Hydrogen

To study the hyperfine structure of Hydrogen, we start with the Maxwell's equations,

$$\vec{\nabla}\cdot\vec{E} = 4\pi\rho,$$

$$\vec{\nabla}\times\vec{B} = \frac{1}{c}\dot{\vec{E}} + \frac{4\pi}{c}\vec{j},$$

$$\vec{\nabla}\times\vec{E} = -\frac{1}{c}\dot{\vec{B}},$$

$$\vec{\nabla}\cdot\vec{B} = 0.$$

They are derived from the action,

$$S = \int dtd^3x\left[\frac{1}{8\pi}\left(\vec{E}^2 - \vec{B}^2\right) - \phi\rho + \frac{1}{c}\vec{A}.\vec{\jmath}\right].$$

A magnetic moment couples to the magnetic field with the Hamiltonian $H = -\vec{\mu}\cdot\vec{B}$, and therefore appears in the Lagrangian as $L = +\vec{\mu}\cdot\vec{B}$. We add this term to the above action,

$$S\int dtd^3x\left[\tfrac{1}{8\pi}\{\vec{E}^2 = \vec{B}^2) - \phi\rho + \tfrac{1}{c}\vec{A}\cdot\vec{j} + \vec{\mu}\cdot\vec{B}\delta(\vec{x}-\vec{y})\right],$$

where $\vec{y}$ is the position of the magnetic moment. The equation of motion for the vector potential is obtained by varying the action with respect to $\vec{A}$.

$$\vec{\nabla}\times\vec{B} = \frac{1}{c}\dot{\vec{E}} + \frac{4\pi}{c}\vec{\jmath} - 4\pi\mu\times\vec{\nabla}\delta\left(\vec{x}-\vec{y}\right).$$

In the absence of time-varying electric field or electric current, the equation is simply,

$$\vec{\nabla}\times\vec{B} = -4\pi\mu\times\vec{\nabla}\delta\left(\vec{x}-\vec{y}\right).$$

It is tempting to solve it immediately as,

$$\vec{B} = -\vec{\mu}\delta(\vec{x}-\vec{y}),$$

but this misses possible terms of the form $\vec{B} \propto \vec{\nabla} f$ where f is a scalar function. To solve it, we use Coulomb gauge and write equation $\vec{\nabla}\times\vec{B} = -4\pi\mu\times\vec{\nabla}\delta(\vec{x}-\vec{y})$ as,

$$-\Delta\vec{A} = -4\pi\mu\times\vec{\nabla}\delta(\vec{x}-\vec{y}).$$

Because $\Delta\frac{1}{|\vec{x}-\vec{y}|} = -4\pi\,\delta(\vec{x}-\vec{y})$ we find,

$$\vec{A}(\vec{x}) = -\mu\times\vec{\nabla}\frac{1}{|\vec{x}-\vec{y}|} = \vec{\mu}\times\frac{\vec{x}-\vec{y}}{|\vec{x}-\vec{y}|^2}$$

The magnetic field is its curl,

$$\vec{B}(\vec{x}) = \vec{\nabla}\times\vec{A} = \vec{\mu}\Delta\frac{1}{|\vec{x}-\vec{y}|} + \vec{\nabla}(\vec{\mu}\cdot\vec{\nabla})\frac{1}{|\vec{x}-\vec{y}|}.$$

We rewrite the latter term as $\nabla_i\nabla_j = \left(\nabla_i\nabla_j - \frac{1}{3}\delta_{ij}\Delta\right) + \frac{1}{3}\delta_{ij}\Delta$, so that the terms in the parenthesis averages out for an isotropic source. They are called the tensor term while the latter the scalar term. Then,

$$\vec{B}(\vec{x}) = \vec{\nabla}\times\vec{A} = -\frac{2}{3}\vec{\mu}\frac{1}{|\vec{x}-\vec{y}|} + \left[\vec{\nabla}(\vec{\mu}\cdot\vec{\nabla})\frac{1}{3}\vec{\mu}\Delta\right]\frac{1}{|\vec{x}-\vec{y}|}$$

After performing differentiation we find,

$$\vec{B}(\vec{x}) = \frac{8\pi}{3}\vec{\mu}\delta(\vec{x}-\vec{y}) + \frac{1}{r^3}\left[3\frac{\vec{r}}{r}\frac{\vec{\mu}.\vec{r}}{r} - \vec{\mu}\right],$$

where we used the notation $\vec{r} = \vec{x}-\vec{y}$.

Finally the interaction of two magnetic moments, $\vec{\mu}_1$ at $\vec{x}$ and $\vec{\mu}_2$ at $\vec{y}$ is given by the magnetic field $\vec{B}(\vec{x})$ reated by the second magnetic moment at $\vec{y}$,

$$H = -\vec{\mu}_1.\vec{B}(\vec{x}) = \frac{8\pi}{3}\vec{\mu}_1.\vec{\mu}_2\delta(\vec{x}-\vec{y}) - \frac{1}{r^3}\left[3\frac{\vec{\mu}_1.\vec{r}}{r}\frac{\vec{\mu}_2.\vec{r}}{r} - \vec{\mu}_1.\vec{\mu}_2\right].$$

In the MKSA system it is,

$$H = -\vec{\mu}_1.\vec{B}(\vec{x}) = -\frac{2\mu_0}{3}\vec{\mu}_1.\vec{\mu}_2\delta(\vec{x}-\vec{y})\frac{\mu_0}{4\pi}\frac{1}{r^3}\left[3\frac{\vec{\mu}_1.\vec{r}}{r}\frac{\vec{\mu}_2.\vec{r}}{r} - \vec{\mu}_1.\vec{\mu}_2\right].$$

For hyperfine splittings in the 1s state of the hydrogen atom Z = 1, the second term vanishes because it is a spherical tensor with q = 2, and hence only the first term is needed. The magnetic moments are (in MKSA),

$$\vec{\mu}_e = g_e \frac{e}{2m_e}\vec{s}_e, \qquad \vec{\mu}_p = g_e \frac{|e|}{2m_p}\vec{s}_p,$$

where $g_e = 2$ and $g_p = 2.79 \times 2$. It is useful to define $\vec{\mu}_e = \frac{|e|\hbar}{2m_e}$ and $\mu_N = \frac{|e|\hbar}{2m_p}$ and

$$\vec{\mu}_e = -\mu_e \frac{2\vec{s}_e}{\hbar}, \quad \vec{\mu}_e = 2.79\mu_N \frac{2\vec{s}_p}{\hbar}.$$

Therefore the Hamiltonian is,

$$H = +\frac{2\mu_0}{3} 2.79\mu_N\mu_e \frac{4}{\hbar^2}\left(\vec{s}_p . \vec{s}_e\right)\delta\left(\vec{x}\right).$$

The first order perturbation of this Hamiltonian gives the hyperfine splitting,

$$E_{hf} = +\frac{2\mu_0}{3} 2.79\mu_N\mu_e \frac{4}{\hbar^2}\left(\vec{s}_p . \vec{s}_e\right)\left|\psi\left(0\right)\right|^2,$$

with $\left|\psi\left(0\right)\right|^2 = \frac{1}{\pi}a_0^{-3}$ for the 1s state. Finally, the eigenvalues of the spin operators are,

$$\vec{s}_p . \vec{s}_e = \frac{1}{2}\left(\left(\vec{s}_p + \vec{s}_e\right)^2 - \vec{s}_p^{\,2} - \vec{s}_e^{\,2}\right) = \begin{cases} \frac{\hbar^2}{4} & (F = 1) \\ -\frac{3\hbar^2}{4} & (F = 0) \end{cases}$$

Therefore the difference in energies is,

$$\Delta E = \frac{2\mu_0}{3} 2.79\mu_N\mu_e \frac{4}{\hbar^2}\left(\frac{\hbar^2}{4} - \frac{-3\hbar^2}{4}\right)\frac{1}{\pi a_0^3}$$
$$= \frac{2\mu_0}{3} 2.79\mu_N\mu_e \frac{4}{\pi a_0^3} = \frac{2\mu_0}{3} 2.79 \frac{e\hbar}{2m_e}\frac{e\hbar}{2m_p}\frac{4}{\pi a_0^3} = 9.39 \times 10^{-25}\ \text{J}.$$

Parametrically, it is $\alpha^2\left(m_e / m_p\right)$ times the binding energy, and hence even more suppressed than the fine structure.

The cosmic thermal bath has T = 2.7 K and hence kT = 3.7×10^{-23} J, which is much larger than the hyperfine splitting.

Einstein Coefficients

In 1917, Einstein introduced *A* and *B* coefficients to describe spontaneous emission and induced absorption and emission. The Einstein *A* coefficient is defined in terms of the total rate of spontaneous emission W_{21}^s from an upper level 2 to a lower level 1 for a system of N_2 atoms in the upper level:

$$W_{21}^s = A_{21}N_2.$$

If level 2 can decay only by radiative emission to level 1, then A_{21} must be the reciprocal, of the spontaneous radiative lifetime t_{spon} of level 2:

$$A_{21} = 1/t_{spon}.$$

(If level 2 can decay to several lower levels, the more general relation $1/t_{spon} = \sum_i A_{2i}$ must be used, where the sum is over all energy levels to which level 2 can decay.)

The B coefficients are defined in terms of the transition rates for (induced) absorption W_{12}^i and induced (or stimulated) emission W_{21}^i:

$$W_{12}^i = B_{12}^\omega \rho\omega N_1$$
$$W_{12}^i = B_{21}^\omega \rho\omega N_2,$$

Where $\rho\omega$ is the energy density per unit angular frequency interval in the region containing N_2 atoms in the upper level and N_1 atoms in the lower level. ρ_ω is assumed to be constant over the frequency range of significant absorption and emission for the 1 ↔ 2 transition. The *B*s have dimensions of volume × angular frequency/(energy × time).

Einstein showed that quite generally,

$$B_{21}^\omega = \left(\pi^2 c^3 / \hbar\omega_{21}^3\right) A_{21}$$

and

$$B_{12}^\omega = \left(g_2 / g_1\right) B_{21}^\omega$$

where ω_{21} is the resonance frequency of the transition. g_1 and g_2 are the degeneracy factors of the two levels. As usual, $\hbar$ is Planck's constant divided by 2π and c is the speed of light. Note that the *B*s, so defined, are independent of the details of the line shape $g(\omega)$ because $\rho\omega$ is assumed to be spectrally flat over the region in which $g(\omega)$ varies significantly.

It is important to recognize that a different relation between the *A* and *B* coefficients is found if some other measure of the radiation energy density is used. For example, if we had used ρ_f (energy density per unit frequency interval), then since $B_{21}^\omega \rho\omega = B_{21}^f \rho_f$ and $\rho_\omega d\omega = \rho_f df$, we find that $B_{21}^f = B_{21}^\omega / 2\pi$. We use superscripts to distinguish the resulting *B*s.

To illustrate the differences that occur in the literature, three examples of the relationships of the Einstein A and B coefficients are quoted.

$$B_{21}^{v} = A_{21} / \left(8\pi hcv_{21}^{3}\right),$$

where $V_{21} = \omega / (2\pi c)$ is the wavenumber of the transition.

$$B_{21}^{f} = c^{3} A_{21} / \left(8\pi hf^{3}\right),$$

where $f = \omega_{21} / 2\pi$. (We do not put subscripts on *f* to avoid possible confusion with socalled *f*-values, to be introduced below.) And finally we have,

$$B_{21}^{f} = c^{2} A_{21} / \left(2hf^{3}\right)$$

One difference among the derivations of these relations is obvious: Herzberg defines *B* in terms of radiation density per unit wavenumber interval while Yariv and Mihalas use radiation density per unit frequency interval. That difference accounts for a factor of c^3 (since $n = f / c$) between Herzberg and Yariv. The remaining factor of *c* difference between Herzberg and Yariv arises because Herzberg defines *B* in terms of irradiance (power per unit area and per unit wavenumber) incident on the atom, while Yariv uses energy per unit volume and per unit frequency interval at the location of the atom. The factor of $c/ 4\pi$ difference between Mihalas and Yariv arises because Mihalas has extracted a factor of 4π in the definition of his *B* in terms of "specific intensity" (power per area and per solid angle and per frequency interval).

The lesson to be learned here is that before one can make use of the formulas for A and B coefficients from a particular source, one must carefully determine which measure of radiation intensity has been used. In fact, in the references cited above the same word "intensity" is used to signify three quite distinct physical quantities.

Relationship between the Einstein coefficients and the absorption cross section: Another pitfall lies in trying to apply the *B* coefficients directly to the analysis of the behavior of nearly monochromatic, directed beams of light. Since most texts provide an inadequate treatment of this important application.

To relate the *B* coefficients to the absorption cross section (or alternatively, to the absorption coefficient), we need to define *B* coefficients for monochromatic and for directed radiation. The careful reader will have already noted that the *B* coefficients were defined above in terms of a broadband, isotropic radiation field. We first consider the case of a monochromatic (but still isotropic) field and define the induced absorption rate due to radiation in the angular frequency range from $\omega + \omega + d\omega$ to be,

$$w_{12}^{i}(\omega)d\omega = b_{12}(\omega)N_{1}\rho(\omega)d\omega,$$

where $\rho(\omega)\ d\omega$ is the energy per unit volume in the range $\omega + \omega + d\omega$. [We recognize that $\rho(\omega)$ is the energy per unit volume and per unit angular frequency interval at ω.]

First, we find the connection between $b_{12}(\omega)$ and B_{12}^{ω} defined previously by letting $\rho(\omega) = \rho_{\omega}$ be a constant over the frequency range near ω_{21}. Then we integrate equation $w_{12}^{i}(\omega)d\omega = b_{12}(\omega)N_1\rho(\omega)d\omega$, over frequency to obtain,

$$W_{12}^{i} = \int_{-\infty}^{+\infty} w_{12}^{i}(\omega)d\omega = \int_{-\infty}^{+\infty} b_{12}(\omega)N_1\rho(\omega)d\omega$$

$$= \rho_\omega N_1\int_{-\infty}^{+\infty} b_{12}(\omega)d\omega = \rho_\omega N_1 B_{12}^{\omega}.$$

To incorporate the atomic frequency response explicitly, we may write,

$$b_{12}(\omega) = B_{12}^{\omega}g(\omega).$$

We now turn to the case of directional radiation. In many practical situations, in particular when dealing with radiation transfer or lasers, it is useful to express the transition rates in terms of the irradiance (time-averaged power per unit area) of a directional beam of light. Classical and quantum-mechanical calculations show that for electric dipole transitions the absorption and stimulated emission rates depend only on the square of the amplitude of the electric field at the location of the atom (and of course on the polarization and frequency spectrum of the light). Hence, as long as the directional beam produces the same energy density (proportional to the electric field amplitude squared) at the location of the atom as does the isotropic field, the transition rate will be the same (taking polarization into account).

The irradiance I is related to the electric field amplitude E by,

$$\mathrm{I} = \frac{1}{2}c\varepsilon_o E^2,$$

where ε_o, as usual, is the permittivity of free space.

For a nearly monochromatic directional beam, the irradiance can be expressed in terms of an integral of $(\rho\omega)$, the energy density in the angular frequency interval between ω and $\omega + d\omega$;

$$I = \int c\rho(\omega)d\omega = \int i(\omega)d\omega,$$

where $i(\omega)$ is the "spectral irradiance" (power per unit area and per unit angular frequency interval). Then the absorption rate due to radiation in the angular frequency range ω to $\omega + d\omega$; is,

$$w_{12}^{i}(\omega)d\omega = N_1 b_{12}(\omega)i(\omega)d\omega / c$$

We can now relate the B_{12}^{ω} coefficient to the absorption cross section $\sigma_a(\omega)$ by the following argument: Using equation $b_{12}(\omega) = B_{12}^{\omega}g(\omega)$ in equation $w_{12}^{i}(\omega)d\omega = b_{12}(\omega)N_1\rho(\omega)d\omega$, we find that power absorbed in the frequency range ω to $\omega + d\omega$; by N_1 atoms is $\hbar\omega B_{12}^{\omega}g(\omega)\rho(\omega)N_1 d\omega$. Suppose that $\rho(\omega)$ is due to a beam of cross-sectional area A. Then $-\Delta P = \hbar\omega B_{12}^{\omega}g(\omega)n_1 Ad\,\omega\Delta x$ is the power lost from this beam as it propagates a distance Δx, where n_1 is the number of atoms per unit volume. The spectral irradiance in the beam is $i(\omega) = c\rho(\omega)$. With the help of,

$$\frac{\Delta P}{A\Delta x d\omega} \rightarrow \frac{di}{dx},$$

we find that,

$$\frac{1}{i(\omega)}\frac{di(\omega)}{dx} = -\hbar\omega n_1 B_{12}^{\omega} g(\omega)/c.$$

Thus the following expressions relate the absorption cross section to the Einstein B coefficient:

$$\sigma_a(\omega) = \hbar\omega B_{12}^{\omega} g(\omega)/c$$
$$\sigma_0 = \hbar\omega_{21} B_{12}^{\omega}/c.$$

In arriving at equation $\sigma_0 = \hbar\omega_{21} B_{12}^{\omega}/c$, we have assumed that $g(\omega)$ is sharply peaked at ω_{21}, and hence that we may replace ω by ω_{21} when carrying out the integration over frequency.

We now consider the absorption process from the point of view of photons. Let R_{12} denote the number of absorption events per unit time and per photon of frequency ω. Then we may find the relationship between σ_a and B_{12}^{ω} by using the standard expression relating the absorption rate per photon to the number density of absorbing atoms n_1 and the relative speed c of the two collision partners:

$$R_{12} = n_1 c\sigma_a(\omega)$$

If we now multiply R_{12} by the number of photons $dN_p(\omega)$ in a volume V in the frequency range ω to $\omega + d\omega$; [$dN_p(\omega)$ is proportional to the energy in that frequency range], we find, with the aid of $\rho(\omega)d\omega = dN_p(\omega)\hbar\omega/V$ that,

$$R_{12}dN_p(\omega) = n_1 c\sigma_a(\omega)\rho(\omega)d\omega V/\hbar\omega$$
$$= N_1 c\sigma_a(\omega)\rho(\omega)d\omega/\hbar\omega$$

If we compare this result with equation $w_{12}^{i}(\omega)d\omega = N_1 b_{12}(\omega)i(\omega)d\omega/c$, we find the relation given in equation $\sigma_a(\omega) = \hbar\omega B_{12}^{\omega} g(\omega)/c$.

Using equation $B_{21}^{\omega} = (\pi^2 c^3/\hbar\omega_{21}^3)A_{21}$ (and $\omega_{12} = 2\pi c/\lambda_{12}$), we may write the cross sections in terms of the A coefficient:

$$\sigma_a(\omega) = \frac{1}{4}(g_2/g_1)\lambda_{21}^2 g(\omega)A_{21},$$

$$\sigma_0 = \frac{1}{4}(g_2/g_1)\lambda_{21}^2 A_{21}.$$

For an electric dipole ("allowed") transition for a stationary, isolated atom, $g(\omega_{21})A_{21}$ is often on the order of unity. Hence, the line center absorption cross section is on the order of λ_{21}^2. We are led to picture the "collision" between a photon and an atom as a collision between a fuzzy ball (the photon) of radius about equal to λ_{21}, and an atom that is small compared to λ_{21}.

For other multipole transitions, for example, for magnetic dipole and electric quadrupole transitions, the product $g(\omega_{21})A_{21}$ may be much smaller than unity if the upper level can decay via electric dipole transitions to other levels. Hence, the cross section will be correspondingly smaller. Obviously, for those other multipole transitions, the fuzzy ball picture of the photon is not appropriate.

Oscillator Strength (F Value)

Oscillator strengths (*f* values) may be defined by comparing the emission rate or absorption rate of the atom with the emission or absorption rate of a classical, singleelectron oscillator (with oscillation frequency ω_{21}). We define an emission oscillator strength f_{21} by the relation,

$$f_{21} = -- A_{21} / \ cl,$$

where,

$$\gamma_{cl} = e^2\omega_{21}^2 / \left(6\pi\varepsilon_o mc^3\right).$$

We have used subscripts on the oscillator strength f_{21} to distinguish it from the transition frequency. Here, *m* is the mass of the electron. The classical radiative decay rate of the single-electron oscillator at frequency ω_{21} is given by γ_{cl}. An absorption oscillator strength f_{12} is then defined by,

$$g_1 f_{12} \equiv -g_2 f_{21} \equiv gf.$$

The *f*s have been defined so that if: (a) $g_2 = 3$ (that is, the angular momentum quantum number J_2 of the upper level is equal to unity), (b) $g_1 = 1$ (that is, $J_1 = 0$), and (c) the Einstein *A* coefficient is equal to the classical decay rate ($A_{21} = \gamma_{cl}$), then the resulting absorption *f* value f_{12} is equal to unity and $f_{21} = -1/3$. Tables of *gf* values for many atomic transitions have been compiled. We may now relate the absorption oscillator strength to the *A* value:

$$f_{12} = \left(g_2 / g_1\right) 2\pi\varepsilon_o mc^3 A_{21} / \left(\omega_{21}^2 e^2\right).$$

Alternatively, we may define the absorption oscillator strength by comparing the absorption cross section of a classical oscillator with that determined by the *B* coefficients. For a stationary, classical oscillator, the absorption cross section is,

$$\sigma_{ac}(\omega) = \frac{\gamma_{cl}/(2\pi)}{(\omega-\omega_o)^2 + (\gamma_{cl}/2)^2} \frac{\pi e^2}{2\varepsilon_o mc}.$$

Note that γ_{cl} is the full-width-at-half-maximum of the absorption curve. With the aid of equation $\sigma_a(\omega) = \hbar\omega B_{12}^{\omega} g(\omega)/c$, we define the absorption oscillator strength f_{12} by the expression,

$$f_{12} = \sigma_0 / \sigma_{0c} = \hbar\omega_{21} B_{12}^{\omega} / \left(c\sigma_{0c}\right),$$

where,

$$\sigma_{0c} = \int_{-\infty}^{+\infty} \sigma_{ac}(\omega) d\omega.$$

Inserting the result stated in equation $\sigma_{ac}(\omega) = \frac{\gamma_{cl}/(2\pi)}{(\omega-\omega_o)^2 + (\gamma_{cl}/2)^2} \frac{\pi e^2}{2\varepsilon_o mc}$ into equation $\sigma_{0c} = \int_{-\infty}^{+\infty} \sigma_{ac}(\omega) d\omega$, we find with the aid of equations $B_{21}^{\omega} = (\pi^2 c^3 / \hbar\omega_{21}^3) A_{21}$ and $f_{12} = (g_2/g_1) 2\pi\varepsilon_o mc^3 A_{21} / (\omega_{21}^2 e^2)$.

Transition Dipole Moment and Line Strength

From a quantum electrodynamics treatment of spontaneous emission, it may be shown that,

$$A_{21} = \frac{2e^2\omega_{21}^3}{3\varepsilon_o hc^3} \sum_{m_1} \left| \langle 1m_1 | \vec{r} | 2m_2 \rangle \right|^2,$$

for a transition from sublevel m_2 of the upper level 2 to all possible m_1 sublevels of the lower level 1. (The usual approximations leading to the electric dipole form of the transition moment have been made. $\vec{r}$ stands for the sum of the electrons' position vectors.) Note that,

$$\sum_{m_1} \left| \langle 1m_1 | \vec{r} | 2m_2 \rangle \right|^2$$

must be independent of m_2. Otherwise, the different m_2 levels would have different lifetimes, which is not possible in an isotropic environment.

If we have a nondegenerate two-state atom, there is only one m_1 and one m_2 and we may unambiguously define the square of the transition dipole moment m_{21} by the relation,

$$e^2 \left| \langle 1m_1 | \vec{r} | 2m_2 \rangle \right|^2 \equiv e^2 r_{21}^2 \equiv \mu_{21}^2$$

If the lower level is degenerate, we then define,

$$e^2 \sum_{m_1} \left| \langle 1m_1 | \vec{r} | 2m_2 \rangle \right|^2 \equiv e^2 r_{21}^2 \equiv \mu_{21}^2$$

as the square of the transition dipole moment. Note again that this moment is independent of m_2. In either case we find,

$$\mu_{21}^2 = A_{21} \frac{3\varepsilon_o hc^3}{2\omega_{21}^3}$$

The "line strength" S_{21} of a transition is defined by the following expression, which is symmetrical in the upper and lower state labels:

$$S_{21} = S_{12} \sum_{m_1 \cdot m_2} \left| \langle 1m_1 | \vec{r} | 2m_2 \rangle \right|^2$$

$$= g_2 e^2 r_{21}^2 = g_2 \mu_{21}^2.$$

Hence, S_{21} is related to A_{21} by,

$$S_{21} = \frac{3\varepsilon_o hc^3}{2\omega_{21}^3} g^2 A_{21}.$$

The line strength S_{12} is the same as the absolute-value-squared of the "reduced matrix element" of the electrons' position operators. In the quantum theory of angular momentum, the reduced matrix element $\langle 1\|r\|2\rangle$ of a vector operator is defined as,

$$\langle 1m_1|r_q|2m_2\rangle = (-1)^{J_1-m_1}\begin{pmatrix} J_1 & J_2 & 1 \\ m_1 & m_2 & q \end{pmatrix}\langle 1|r|2\rangle$$

where $q = 0, \pm 1$ and J_i is the angular momentum quantum number for the ith level. The factor in parentheses is the Wigner 3-j symbol. The relation with line strength is,

$$S_{12}\left|\langle 1\|r\|2\rangle\right|^2$$

Rabi Frequency

In many applications involving the interaction of laser radiation with atomic and molecular systems, coherent effects are important. For example, the population difference in a two-level system driven by a beam of coherent radiation oscillates in time with an angular frequency called the "Rabi frequency." The more intense the light source, the more rapidly the population difference oscillates. The Rabi frequency for on-resonance excitation (when the frequency ω of exciting light equals the resonance frequency ω_{21}) can be expressed in terms of the electric field amplitude of the linearly polarized light field and the transition dipole moment μ_{21} by the following expression:

$$\omega_R = \mu_{21} E / \hbar.$$

If the upper level and lower level are degenerate, there may be several Rabi frequencies, one for each value of $\langle 1m_1|r|2m_2\rangle$. In that case, the dynamical behavior of the system may be quite complicated.

Some confusion in the definition of the Rabi frequency occurs because many authors make use of the so-called rotating-wave approximation, in which $E\cos(\omega t)$ is replaced by $E\left[e^{i\omega t} + e^{i\omega t}\right]$. Terms with a time dependence $e^{2i\omega t}$ are then dropped from the equations of motion. In terms of E, the Rabi frequency is,

$$\omega_R = \mu_{21} 2E / \hbar$$

Unfortunately, it is not always obvious which form of the electric field amplitude has been adopted. Some authors use the root-mean-square electric field $E_{rms} = E / \sqrt{2}$.

Interaction of an Atom with Radiation

Every object in the universe is made up of atoms. Atoms are made up of extremely small particles such as electrons, protons, and neutrons. Electrons are the negatively charged particles and protons are the positively charged particles. Neutrons have no charge. Hence, neutrons are referred as neutral particles.

The strong nuclear force between the protons and neutrons makes them stick together to form the nucleus. Neutrons have no charge. so the overall charge of the nucleus is positive because of the protons.

The electrostatic force of attraction between the nucleus and electrons causes electrons to revolve around the nucleus.

The electrons revolving around the nucleus have different energy levels based on the distance from the nucleus.

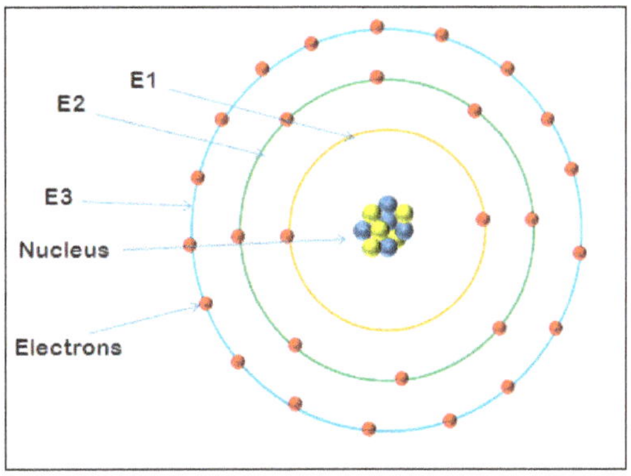

The electrons revolving very close to the nucleus have lowest energy level whereas the electrons revolving at the farthest distance from nucleus have highest energy level.

For example, if the lowest energy level is E_1 the next energy level is E_2 and next is E_3, E_4 and so on.

The electrons in the lower energy state (E_1) needs extra energy to jump into next higher energy state (E_2). This energy can be supplied in the form of the electric field, heat or light.

When the electrons in the lower energy state (E_1) gains sufficient energy from photons, they jump into next higher energy state (E_2).

The electrons in the higher energy state do not stay for long period. After a short period, they again fall back to the lower energy level by losing their energy. The electrons in the higher energy level or higher energy state lose energy in the form of light before they fall back to the lower energy state.

The electrons in the higher energy state are known as excited electrons whereas the electrons in the lower energy state are known as ground electrons.

In lasers, the way light or photons interact with atoms plays an important role in its operation. The photons interact in three ways with the atoms:

- Absorption of radiation or light.

- Spontaneous emission.
- Stimulated emission.

Absorption of radiation or light

The process of absorbing energy from photons is called absorption of radiation.

It is well known that there are different energy levels in an atom. The electrons that are very close to the nucleus have lowest energy level. These electrons are also known as ground state electrons.

Let us consider that the energy level of ground state electrons or lower energy state electrons is E_1 and the next higher energy level or higher energy state is E_2.

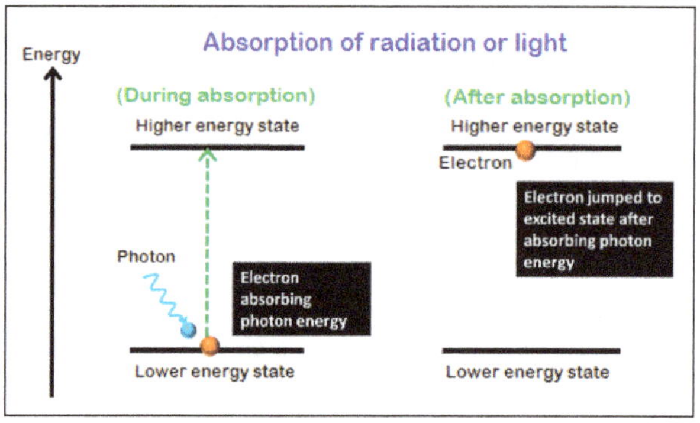

When ground state electrons or lower energy state electrons (E_1) absorbs sufficient energy from photons, they jump into the next higher energy level or higher energy state (E_2). In other words, when the ground state electrons absorb energy which is equal to the energy difference between the two energy states ($E_2 - E_1$), the electrons jumps from ground state (E_1) to the excited state or higher energy level (E_2). The electrons in the higher energy level are called excited electrons.

The light or photons energy applied to excite the electrons can be mathematically written as:

$$hv = E_2 - E_1$$

where,

- h = Planck's constant.
- V = Frequency of photon.
- E_1 = Lower energy level electrons or ground state electrons.
- E_2 = Higher energy level electrons or excited state electrons.

Absorption occurs only if the energy of photon exactly matches the difference in energy between the two electron shells or orbits.

Spontaneous Emission

The process by which excited electrons emit photons while falling to the ground level or lower energy level is called spontaneous emission.

Electrons in the atom absorb energy from various sources such as heat, electric field, or light. When the electrons in the ground state or lower energy state (E_1) absorb sufficient energy from photons, they jump to the excited state or next higher energy state (E_2).

The electrons in the excited state do not stay for a long period because the lifetime of electrons in the higher energy state or excited state is very small, of the order of 10^{-8} sec. Hence, after a short period, they fall back to the ground state by releasing energy in the form of photons or light.

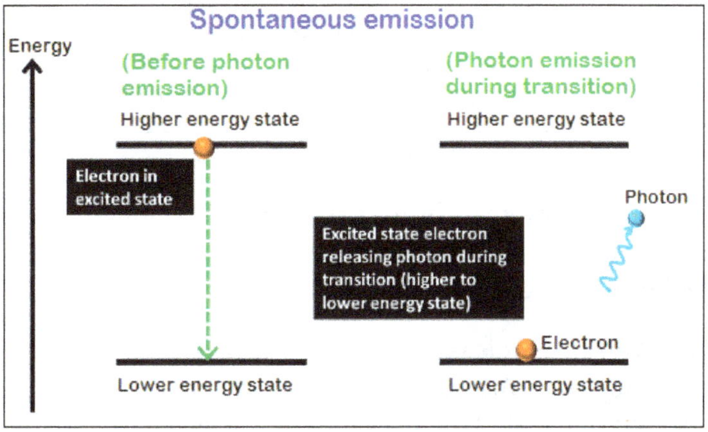

The energy of the emitted photon is directed proportional to the energy gap of the material. The materials with large energy gap will emit high-energy photons or high-intensity light whereas the materials with small energy gap will emit low energy photons or low-intensity light.

The energy of released photon is equal to the difference in energies between the two electron shells or orbits.

The energy of the excited electrons can also be released in other forms such as heat. If the excited state electrons release energy in the form of photons or light while falling to the ground state, the process is called spontaneous emission.

In spontaneous emission, the electrons changing from one state (higher energy state) to another state (lower energy state) occurs naturally. So the photon emission also occurs naturally or spontaneously.

The photons emitted due to spontaneous emission do not flow exactly in the same direction of incident photons. They flow in the random direction.

Stimulated Emission

The process by which electrons in the excited state are stimulated to emit photons while falling to the ground state or lower energy state is called stimulated emission. Unlike the spontaneous emission, in this process, the light energy or photon energy is supplied to the excited electrons instead of supplying energy to the ground state electrons.

The stimulated emission is not a natural process it is an artificial process. In stimulated emission, the electrons in the excited state need not wait for natural spontaneous emission to occur. An alternative method is used to stimulate excited electron to emit photons and fall back to ground state.

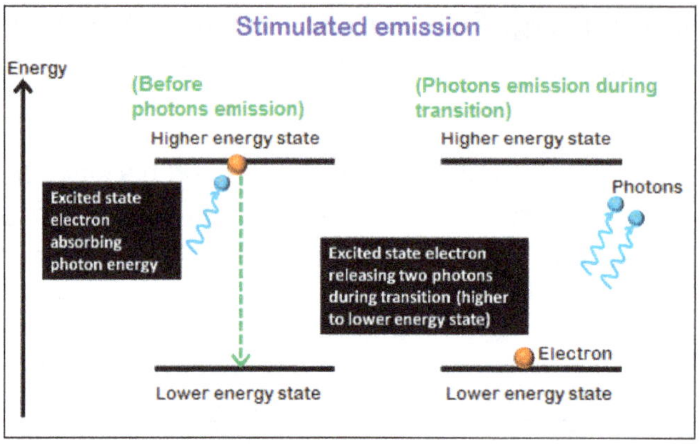

Physics and Radio-Electronics.

The incident photon stimulates or forces the excited electron to emit a photon and fall into a lower state or ground state. The energy of a stimulating or incident photon must be equal to the energy difference between the two electron shells.

In this process, the excited electron releases an additional photon of same energy (same frequency, same phase, and in the same direction) while falling into the lower energy state. Thus, two photons of same energy are released while electrons falling into the ground state. In stimulated emission process, each incident photon generates two photons.

The photons emitted in the stimulated emission process will travel in the same direction of the incident photon. Many ways exist to produce light, but the stimulated emission is the only method known to produce coherent light (beam of photons with the same frequency). All the photons in the stimulated emission have the same frequency and travel in the same direction.

References

- Atomic-spectroscopy: oxinst.com, Retrieved 17 July, 2019
- Atomic-absorption-spectroscopy-aas: chemicool.com, Retrieved 14 April, 2019
- A-Atomic-Absorption-Spectroscopy: libretexts.org, Retrieved 9 January, 2019
- Atomic-Spectra, The-Hydrogen-Atom, Physical-and-Theoretical-Chemistry: libretexts.org, Retrieved 20 June, 2019
- Fine-structure, science: britannica.com, Retrieved 25 February, 2019
- Hyperfine-structure, science: britannica.com, Retrieved 5 May, 2019
- Oscillator-formulas: homepage.univie.ac.at, Retrieved 3 March, 2019

6

Spin–orbit Coupling

The relativistic interaction of a particle's spin with its motion inside a potential is referred to as spin-orbit coupling. Some of the different types of coupling schemes are JJ coupling and RS coupling. The topics elaborated in this chapter will help in gaining a better perspective about these types of spin-orbit coupling.

Spin-orbit coupling refers to the interaction of a particle's "spin" motion with its "orbital" motion.

Spin-orbit Coupling Hamiltonian

The magnitude of spin-orbit coupling splitting is measured spectroscopically as,

$$\begin{aligned} H_{so} &= \frac{1}{2} hcA((l+s)(l+s+1) - l(l+1) - s(s+1)) \\ &= \frac{1}{2} hcA(l^2 + s^2 + ls + sl + l + s - l^2 - l - s^2 - s)) \\ &= hcAl \cdot s \end{aligned}$$

The expression can be modified by realizing that $j = l + s$.

$$H_{so} = \frac{1}{2} hcA\big(j(j+1) - l(l+1) - s(s+1)\big)$$

where A is the magnitude of the spin-orbit coupling in wave numbers. The magnitude of the spin orbit coupling can be calculated in terms of molecule parameters by the substitution,

$$hcA\,\hat{L} \cdot \hat{S} = \frac{Z\alpha^2}{2} \frac{1}{r^3} \hat{L} \cdot \hat{S}$$

where a is the fine structure constant (a =1/137.037) and the carrots indicate that L and S are operators. The fine structure constant is a dimensionless constant, $a = \frac{e^2}{\text{ac}}$. Z is an effective atomic number. The spin orbit coupling splitting can be calculated from,

$$E_{so} = \int \Psi^* H_{so} \Psi dr = \frac{Z}{2(137)^2} \int \Psi^* \frac{\hat{L} \cdot \hat{S}}{r^3} dr$$

This expression can be recast to give an spin-orbit coupling energy in terms of molecular parameters,

$$E_{so} = \frac{1}{2}\left(j(j+1) - l(l+1) - s(s+1)\right) = \frac{Z}{2(137)^2}\left\langle \frac{1}{r^3} \right\rangle$$

where,

$$\left\langle \frac{1}{r^3} \right\rangle = \int \Psi^* \left\langle \frac{1}{r^3} \right\rangle \Psi dr$$

We can evaluate this integral explicitly for a given atomic orbital.

For example for Ψ_{210} we have,

$$\Psi_{210} = \frac{1}{4\sqrt{2\pi}}\left(\frac{Z}{a_0}\right)^{\frac{3}{2}} \frac{Zr}{a_0} e^{-Zr/2a_0} \cos\theta$$

so that the integral is,

$$\left\langle \frac{1}{r^3} \right\rangle = \frac{1}{32\pi}\left(\frac{Z}{a_0}\right)^5 \int_0^{2z} d\phi \int_0^{z} \cos^2\theta \sin\theta \, d\theta \cos\theta \int_0^{\infty} r^2 e^{Zr/a_0} \left\langle \frac{1}{r^3} \right\rangle r^2 dr$$

which integrates to,

$$\left\langle \frac{1}{r^3} \right\rangle = \frac{1}{32\pi}\left(\frac{Z}{a_0}\right)^5 2\pi \left(\frac{2}{3}\right)\left(\frac{a_0^{\,2}}{Z^2}\right) = \frac{1}{24}\left(\frac{Z}{a_0}\right)^3$$

or $Z^3/24$ in atomic units.

Therefore in atomic units we have,

$$\left\langle \frac{1}{r^3} \right\rangle = \frac{Z^3}{n^3 l(l+1/2)(l+1)}$$

Therefore, in general the spin-orbit splitting is given by,

$$E_{so} = \frac{Z^4}{2(137)^2 n^3}\left(\frac{j(j+1) - l(l+1) - s(s+1)}{2l(l+1/2)(l+1)}\right)$$

Note that the spin-orbit coupling increases as the fourth power of the effective nuclear charge Z, but only as the third power of the principal quantum number *n*. This indicates that spin orbit-coupling interactions are significantly larger for atoms that are further down a particular column of the periodic table.

LS Coupling

In atomic spectroscopy, Russell–Saunders coupling, also known as LS coupling, specifies a coupling scheme of electronic spin- and orbital-angular momenta. The coupling scheme is named after H. N. Russell and F. A. Saunders. Russell-Saunders coupling is useful mainly for the lighter atoms, roughly for atoms with atomic number less than 57. For heavier atoms j-j coupling gives a better approximation for atomic wave function.

Consider an atom with n electrons. In Russell-Saunders coupling the orbital angular momentum eigenstates of these electrons are coupled to eigenstates with quantum number L of the total angular momentum operator squared L^2, where the angular momentum operator is:

$$\mathrm{L} \equiv \sum_{i=1}^{n} \mathrm{l}(i).$$

Separately the one-electron spin functions are coupled to eigenstates with quantum number S of total spin angular momentum squared S^2. Sometimes there is further coupling to $\mathrm{J} \equiv \mathrm{L} + \mathrm{S}$. The resulting L-S eigenstates are characterized by term symbols.

Consider, as an example, the excited helium atom in the atomic electron configuration $2p3p$. By the triangular conditions the one-electron spins $s = ½$ can couple to $|½-½|$, $½+½ = 0$, 1 (spin singlet and triplet) and the two orbital angular momenta $l = 1$ can couple to $\mathrm{L} = |1-1|$, 1, $1+1 = 0$, 1, 2. In total, Russell-Saunders coupling gives two-electron states labeled by the term symbols: ^{1}S, ^{1}P, ^{1}D, ^{3}S, ^{3}P, ^{3}D.

The dimension is $1\times(1+3+5) + 3\times(1+3+5) = 36$. The electronic configuration $2p3p$ stands for $6\times6 = 36$ orbital products, as each of the three p-orbitals has two spin functions, so that in total there are 6 spinorbitals with principal quantum number $n = 2$ and also 6 spinorbitals with $n = 3$. A check on dimensions before and after coupling is useful because it is easy to overlook coupled states.

Russell-Saunders coupling gives useful first-order states in the case that one-electron spin-orbit coupling is much less important than the Coulomb interactions between the electrons and can be neglected. This occurs for the low Z part (i.e., the lighter elements) of the periodic table, roughly up to the lanthanoids (previously lanthanides, starting at $Z = 57$). The usefulness stems from the fact that states of different L and S do not mix under the total Coulomb interaction, so that LS coupling achieves a considerable block diagonalization of the matrix of a Hamiltonian in which spin-orbit coupling is absent.

In the high Z regions of the periodic system it is common to first couple the one-electron momenta $\mathrm{j} \equiv \mathrm{l} + \mathrm{s}$ and then the one-electron j-eigenstates to total J. This so-called j-j coupling scheme gives a more useful first-order approximation when spin-orbit interaction is larger than the Coulomb interaction and spin-orbit interaction is included, while the Coulomb interaction is neglected. If, however, in either coupling scheme all resulting states are accounted for, i.e., the same subspace of Hilbert (function) space is obtained, then the choice of coupling scheme is irrelevant in calculations where both interactions—electrostatic and spin-orbit—are included on equal footing.

More Complicated Electron Configurations

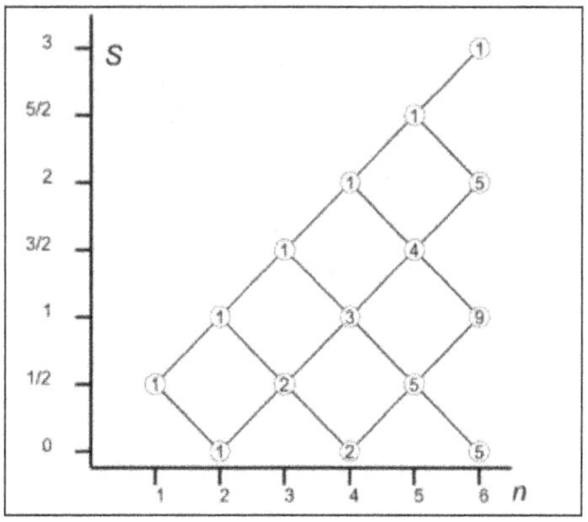

Spin branching diagram. Number of electrons n on the abscissa. Spin quantum number S on the ordinate.

Spin Coupling

Let us consider how to couple n spin-½ particles (electrons) to eigenstates of total S^2.

The one-electron system ($n = 1$) has two functions with $s = ½$ and $m = \pm ½$ (spin up and down). It was shown that addition of an electron leads to four two-electron spin functions: a singlet ($2S + 1 = 1$) and a triplet space (a "ladder" of $2S + 1 = 3$ spin functions). If one adds one spin to the two-electron triplet ($S = 1$), the triangular conditions tell us that $S = 3/2$, and $S = 1/2$ can be obtained. In the diagram above we see that two lines depart from the $n = 2$, $S = 1$ node. One line goes up and joins the $n = 3$, $S = 3/2$ node, the other goes down and joins the $n = 3$, $S = 1/2$ node.

A two-particle spin system also has a spin-singlet $S = 0$. Adding a spin to it leads to a doublet, $S = 1/2$. It can be shown that the doublet thus obtained is orthogonal to the doublet with the intermediate ($n = 2$, $S = 1$) spin triplet. Two different paths lead two different—even orthogonal—three-electron spin doublets. This is why the number 2 is listed in the circle at the $n = 3$, $S = 1/2$ node.

From the two doublets it is possible, by again adding one electron, to create two four-electron singlets, and two four-electron triplet spaces (ladders). A four-electron triplet space can also be obtained from the three-electron quadruplet ($S = 3/2$) by subtracting $S = 1/2$. Hence the number 3 is found at the $n = 4$, $S = 1$ node. It is the sum of numbers in the nodes directly on its left connected to it.

The system of sequentially coupling one electron at the time should be clear now. Every path in the branching diagram corresponds to a unique ($2S+1$)-dimensional ladder of eigenfunctions of total spin angular momentum operator S^2 with spin quantum number S. Ladders belonging to different paths are orthogonal. The number in the circle indicates how many paths have this node as endpoint.

It is of interest to check dimensions. For instance, the total spin space of five electrons is of dimension $2^5 = 32$. One finds $5\times2 + 4\times4 + 1\times6$ (second factors being $2S+1$), which indeed adds up to 32.

The actual coupling of electron n to a state of $n-1$ electrons uses special values of Clebsch-Gordan coefficients. Let us use k as a path index (e.g., for $S = 1$ and $n = 6$, the index k runs from 1 to 9). Then addition (line upwards) gives the n-electron spin state:

$$|k,S,M\rangle = \sqrt{\frac{1}{2S}}\left(\sqrt{S+M}\,|k,S-\tfrac{1}{2},M-\tfrac{1}{2}\rangle\,|\tfrac{1}{2},\tfrac{1}{2}\rangle + \sqrt{S-M}\,|k,S-\tfrac{1}{2},M+\tfrac{1}{2}\rangle\,|\tfrac{1}{2},-\tfrac{1}{2}\rangle\right).$$

Subtraction (line downwards) gives the n-electron spin state:

$$|k,S,M\rangle = \sqrt{\frac{1}{2S+2}}\left(-\sqrt{S-M+1}\,|k,S+\tfrac{1}{2},M-\tfrac{1}{2}\rangle\,|\tfrac{1}{2},\tfrac{1}{2}\rangle + \sqrt{S+M+1}\,|k,S+\tfrac{1}{2},M+\tfrac{1}{2}\rangle\,|\tfrac{1}{2},-\tfrac{1}{2}\rangle\right)$$

In particular, for $n = 2$ the spin triplet results:

$$|1,1,M\rangle = \sqrt{\frac{1}{2}}\left(\sqrt{1+M}\,|1,\tfrac{1}{2},M-\tfrac{1}{2}\rangle\,|\tfrac{1}{2},\tfrac{1}{2}\rangle + \sqrt{1-M}\,|1,\tfrac{1}{2},M+\tfrac{1}{2}\rangle\,|\tfrac{1}{2},-\tfrac{1}{2}\rangle\right),$$

which gives the triplet "ladder", where in the notation the index $k = 1$ is suppressed,

$$|1,1\rangle = |\tfrac{1}{2},\tfrac{1}{2}\rangle\,|\tfrac{1}{2},\tfrac{1}{2}\rangle \equiv \alpha(1)\alpha(2)$$
$$|1,0\rangle = \sqrt{\tfrac{1}{2}}\left(|\tfrac{1}{2},-\tfrac{1}{2}\rangle\,|\tfrac{1}{2},\tfrac{1}{2}\rangle + |\tfrac{1}{2},\tfrac{1}{2}\rangle\,|\tfrac{1}{2},-\tfrac{1}{2}\rangle\right) \equiv \sqrt{\tfrac{1}{2}}\left(\beta(1)\alpha(2)+\alpha(1)\beta(2)\right)$$
$$|1,-1\rangle = |1,\tfrac{1}{2},-\tfrac{1}{2}\rangle\,|\tfrac{1}{2},-\tfrac{1}{2}\rangle \equiv \beta(1)\beta(2)$$

Here the more common notation for one-electron spin functions (α for $m = ½$ and β for $m = -½$) is introduced. Note that all three triplet spin functions are symmetric under interchange of the two electrons.

The two-electron spin singlet follows from (suppressing the path index $k = 1$ again):

$$|0,0\rangle = \sqrt{\tfrac{1}{2}}\left(-|\tfrac{1}{2},-\tfrac{1}{2}\rangle\,|\tfrac{1}{2},\tfrac{1}{2}\rangle + |\tfrac{1}{2},\tfrac{1}{2}\rangle\,|\tfrac{1}{2},-\tfrac{1}{2}\rangle\right) \equiv \sqrt{\tfrac{1}{2}}\left(\alpha(1)\beta(2)-\beta(1)\alpha(2)\right)$$

a function that is antisymmetric (changes sign) under permutation of electron 1 and 2.

Orbital Coupling

The one-electron orbital angular momenta l can also be coupled sequentially, i.e., one after the other. Different paths give orthogonal states. Consider as an example the configuration p^3. Repeated application of the triangular conditions gives that $3^3 = 27$ dimensional space decomposes as follows in eigenspaces of L^2, [with $L \equiv l(1) + l(2) + l(3)$]:

(2)F, (2)D, (1)D, (2)P, (1)P, (0)P, (1)S

where the two-electron quantum numbers L_{12} of electron 1 and 2 precede in brackets the letters designating the three-electron angular momentum quantum number. Dimensions are: 7 + 5 + 5 + 3 + 3 + 3 + 1 = 27. There are two orthogonal D spaces and three orthogonal P spaces, while there is only one F and S space (of dimension 7 and 1, respectively).

The algorithm should be clear now. For instance, to decompose the 135-dimensional space belonging to the configuration p^3d, one can use the result for p^3 and add d to any of the eigenspaces and use the triangular conditions again, as in $F \times d$ gives $L = 5, 4, 3, 2, 1$.

Equivalent Electrons

So far it was ignored the Pauli principle that states that many-electron wave functions must be antisymmetric under simultaneous transposition of spatial and spin coordinates of any pair of electrons. This principle affects considerably the results for configurations of equivalent electrons, which by definition are electrons with the same n (principal) and l (azimuthal or angular momentum) quantum numbers. It will be shown that certain terms (*LS* states) arising from equivalent electrons are forbidden by the Pauli principle. Terms arising from non-equivalent electrons are never forbidden, they simply give rise to non-vanishing antisymmetric states.

Two-electron Atoms

Let us take as an example the spatial product, describing two equivalent electrons, $2p_x(1)2p_x(2)$. Transposition $P_{(12)}$ of space coordinates gives in general:

$$P_{(12)}\, p_i(1)p_j(2) = p_i(2)p_j(1) \equiv p_j(1)p_i(2).$$

Hence,

$$P_{(12)}\, p_x(1)p_x(2) = p_x(1)p_x(2),$$

and it follows that the product is symmetric under transposition.

So, there are two-electron singlet and triplet spin functions. In both cases $M_S = 0$ functions are considered, thus,

$$\begin{aligned} |\, p_x^2, S=0, M_S=0\rangle &\equiv 2p_x(1)2p_x(2)\big(\alpha(1)\beta(2)-\beta(1)\alpha(2)\big)/\sqrt{2} \\ |\, p_x^2, S=1, M_S=0\rangle &\equiv 2p_x(1)2p_x(2)\big(\alpha(1)\beta(2)+\beta(1)\alpha(2)\big)/\sqrt{2}. \end{aligned}$$

Antisymmetric functions, satisfying the Pauli principle, are obtained by use of the antisymmetrizer $\mathcal{A} = [(1)-(12)]/2$, where (12) transposes *simultaneous* space *and* spin coordinates of electron 1 and 2 and (1) is the identity operator (does nothing).

$$\begin{aligned} 2\mathcal{A}\,|\, p_x^2, S=0, M_S=0\rangle &= 2p_x(1)2p_x(2)\big(\alpha(1)\beta(2)-\beta(1)\alpha(2)\big)/\sqrt{2} \\ &\quad -2p_x(2)2p_x(1)\big(\alpha(2)\beta(1)-\beta(2)\alpha(1)\big)/\sqrt{2} \\ &= 2\Big[2p_x(1)2p_x(2)\big(\alpha(1)\beta(2)-\beta(1)\alpha(2)\big)/\sqrt{2}\Big]. \end{aligned}$$

Hence,

$$\mathcal{A}\,|\, p_x^2, S=0, M_S=0\rangle = |\, p_x^2, S=0, M_S=0\rangle.$$

It follows that the symmetric spatial function $2p_x(1)2p_x(2)$ multiplied by the antisymmetric spin singlet function is an eigenfunction of the antisymmetrizer, that is, the symmetric space times antisymmetric spin function satisfies the Pauli principle.

The calculation for the spin triplet is repeated:

$$\begin{aligned} 2\mathcal{A}\,|\,p_x^2, S=1, M_S=0\rangle &= 2p_x(1)2p_x(2)\big(\alpha(1)\beta(2)+\beta(1)\alpha(2)\big)/\sqrt{2} \\ &\quad -2p_x(2)2p_x(1)\big(\alpha(2)\beta(1)+\beta(2)\alpha(1)\big)/\sqrt{2} \\ &= 0 \end{aligned}$$

Hence,

$$\mathcal{A}\,|\,p_x^2, S=1, M_S=0\rangle = 0.$$

The symmetric spatial function $2p_x(1)2p_x(2)$ multiplied by the symmetric spin triplet function does not have a non-vanishing antisymmetric component, that is, this space-spin product is forbidden by the Pauli principle.

So, this example shows clearly two important results valid for two particles:

1. Spin eigenfunctions of S^2 are either symmetric or antisymmetric under transposition of spin coordinates.
2. The symmetric orbital function multiplied by a symmetric spin function is Pauli forbidden, i.e., vanishes upon antisymmetrization. The same function multiplied by an antisymmetric spin function is antisymmetric under simultaneous transpositions of space and spin coordinates, and hence is Pauli allowed.

Conversely, it can be shown that an antisymmetric orbital function, for instance: $p_x(1)\,p_y(2) - p_x(2)\,p_y(1)$, can only combine with a symmetric spin function to a non-vanishing—Pauli allowed—totally antisymmetric spin-orbit function.

In general one can show from the symmetry of the Clebsch-Gordan coefficients that eigenstates of L^2 for two equivalent electrons with odd quantum number L are antisymmetric under transposition of the electron coordinates. Hence these odd L states must be multiplied by symmetric spin functions (triplets $S = 1$). In total $L+S$ must be even. The even L states (still for two equivalent electrons) are even under transposition and must be multiplied by odd spin functions (singlets, $S = 0$). Again $L+S$ must be even in order to get a non-vanishing result upon antisymmetrization.

Before continuing, it must be pointed out that there exists an interesting symmetry between *holes* and particles. If an electron from a closed subshell is removed a *hole* is prepared in the subshell. It can be shown that a hole shares many properties with an electron: it has the same orbital and spin angular momentum quantum number. A number of holes must satisfy the Pauli principle in the very same way as an equal number of electrons.

The rule is: *$L + S$ must be even for two equivalent electrons.* This rule also holds for two equivalent holes. Thus, for instance the carbon atom ground state has two equivalent electrons: $1s^2\ 2s^2$

$2p^2$, which gives ^3P, ^1D and ^1S. The oxygen atom ground state has two equivalent holes: $1s^2\ 2s^2\ 2p^4$, which also gives ^3P, ^1D and ^1S.

Knowing all this one can construct the following table,

$$\begin{aligned} (ns)^2 &\rightarrow\ {}^1S \\ (np)^2, (np)^4 &\rightarrow\ {}^1S, {}^1D, {}^3P \\ (nd)^2, (nd)^8 &\rightarrow\ {}^1S, {}^1D, {}^1G, {}^3P, {}^3F \\ (nf)^2, (nf)^{12} &\rightarrow\ {}^1S, {}^1D, {}^1G, {}^1I, {}^3P, {}^3F, {}^3H \end{aligned}$$

Many-electron Atoms

When considering more than two electrons, one can ignore closed (sub)shells in the Russell-Saunders coupling because they have $L = S = 0$.

For more than two electrons (or holes) the theory based on the symmetric group can be extended. It can be shown that an N-electron spin eigenfunction of S^2 belongs to an irreducible representation of the symmetric group (also known as permutation group) S_N. In other words, the concept of symmetry under transpositions can be generalized to more than two particles. This generalized permutation symmetry is enforced on the spin functions by requiring them to be eigenfunctions of S^2, a Casimir operator of the group SU(2). The irreducible representations of SU(2) and S_N carried by tensor space are intertwined. The antisymmetrizer acting on an orbital product times an eigenfunction of S^2 enforces permutational symmetry on the orbital product (carries, so to speak, the permutation symmetry of the spin part over to the orbital part). The permutational symmetry of the orbital product, finally, implies that the orbital product is adapted to the unitary group $U(2\ell+1)$. The irreducible representations of the latter group carry different eigenstates of L^2 that, by construction, are Pauli allowed. This procedure is elegant and not very tedious, but requires some knowledge of group theory, in particular knowledge of the connection between the irreducible representations of SU(2), S_N and $U(2\ell+1)$ carried by N-fold tensor product spaces.

The conventional, much more tedious, non group-theoretical, way to proceed is by a "bookkeeping" procedure. Basically, one tabulates all the combinations of one-electron quantum numbers allowed by the Pauli principle and determines the combination with highest M_L and M_S, which are given by the respective sums of the orbit- and spin- magnetic quantum numbers of the individual electrons. This combination is unique and defines "ladders" of eigenstates with $L = M_L^{max}$ and $S = M_S^{max}$. One removes the ladders starting with these maximum values from the table and searches for the next unique combination of highest M_L and M_S that are left in the table. This is repeated until the table is empty and all ladders (LS states) have been assigned.

It is of some historical interest to remark that the latter procedure was followed by Hund in 1925, before the discovery of spin or introduction of Slater determinants. The following table is a combination of tables 3 and 4 of Hund with the correction of two "bookkeeping" errors made by Hund, proving that this procedure is error prone.

Term symbols deriving from configurations of equivalent electrons	
Configuration	Terms
p, p[5]	^{2}P
p[2], p[4]	^{1}S, ^{1}D, ^{3}P
p[3]	^{2}P, ^{2}D, ^{4}S
d, d[9]	^{2}D
d[2], d[8]	^{1}S, ^{1}D, ^{1}G, ^{3}P, ^{3}F
d[3], d[7]	^{2}P, ^{2}D, ^{2}D, ^{2}F, ^{2}G, ^{2}H, ^{4}P, ^{4}F
d[4], d[6]	^{1}S, ^{1}S, ^{1}D, ^{1}D, ^{1}F, ^{1}G, ^{1}G, ^{1}I, ^{3}P, ^{3}P, ^{3}D, ^{3}F, ^{3}F, ^{3}G, ^{3}H, ^{5}D
d[5]	^{2}S, ^{2}P, ^{2}D, ^{2}D, ^{2}D, ^{2}F, ^{2}F, ^{2}G, ^{2}G, ^{2}H, ^{2}I, ^{4}P, ^{4}D, ^{4}F, ^{4}G, ^{6}S

JJ Coupling

In the case of atoms with a large number of electrons the terms pro $\vec{l}_i.\vec{s}_i$ are more important than the electrostatic behavior in the mean field. So the latter are considered as a small perturbation. This fact leads to the behavior of the atoms that their electrons first $\vec{l}_i$ and $\vec{s}_i$ to single electron total angular momenta $\vec{j}_i$ and those add up (vectorially) to a whole system total angular momentum $\vec{J}_i$.

$$\vec{j}_i = \vec{l}_i + \vec{s}_i$$

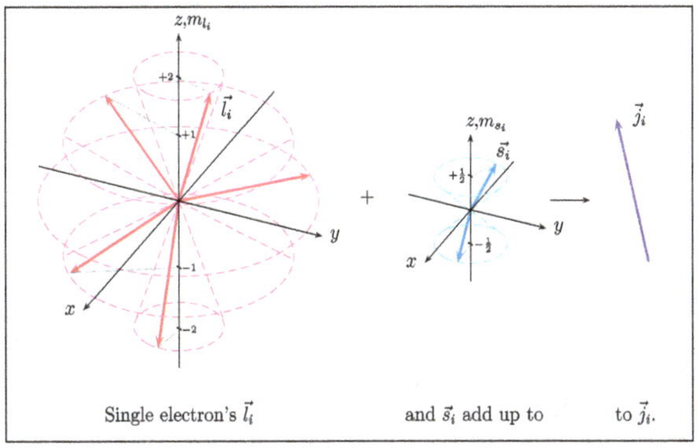

In figure, illustration of the angular momentum vector $\vec{l}_i$ of a single electron adding up with the single electron's spin $\vec{s}_i$ yielding the single electron's total angular momentum $\vec{j}_i$. That this happens "before" those $\vec{j}_i$ are adding up to the whole atom's total angular momentum distinguishes the j j-coupling from the LS-coupling. Again all those vectors are on cones due to the uncertainty principle and their lengths are equal to $\sqrt{l_i(l_i+1)}$, $\sqrt{s_i(s_i+1)}$ and $\sqrt{j_i(j_i+1)}$. All physically possible (Pauli exclusion principle) orientated single electron variable vectors "then" vectorially add up to the whole system total angular momentum $\vec{J}$,

$$\vec{J}_i = \sum_{i=1}^{N} \vec{j}_i$$

The same combinatoric machinery as in the LS-coupling takes place in the j j-case; but the other way around. The single electron vector quantities have some possible m_{li} and m_{si} "Then" they could add up to lots of possible $\vec{j}_i$. Finally those – holding values of m_{ji} – add up to J. But again only those combinations of the single electron quantities that obey the Pauli principle are physically possible. In the case of LS-coupling the equivalent electrons are those with a common n and l, but in the case of j j-coupling the electrons that have a common n_i, l_i, j_i. Also the set of quantum numbers that are important for the coupling are n, l, j, m_j.

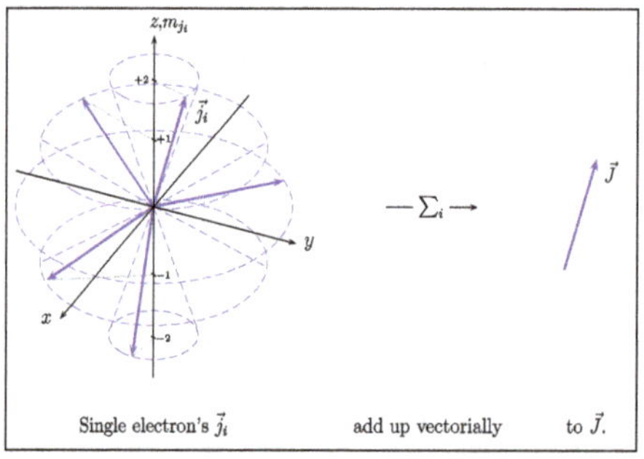

Figure: Illustration of the single electrons angular momentum vector $\vec{j}_i$ for the case of $j_i = 2$. This enables the vector to have a m_{j_i} from −2 up to +2 in steps of the size 1. All those vectors are on cones due to the uncertainty principle and their length is equal to $\sqrt{j_i(j_i+1)}$. All physically possible (Pauli exclusion principle) orientated groups of single electron angular momenta for the whole atom "then" add vectorially up to the total system angular momentum $\vec{J}_i$.

For this coupling scheme there are several conventions to denote the possible states. The following general form was chosen to enable the simple identification of possible transitions between the term symbols.

$$\left(j_1, j_2, \ldots, j_N\right)_J^{(o)}$$

The "∘" in the top right index denotes the odd parity if present just as used in LS coupling nomenclature. To provide a more precise indication of possible states by these terms it is possible to couple electrons in the same shell of equivalent electrons (same n_i, l_i, j_i) first and then couple those shell total angular momenta successively, similarly to the parental history in LS-coupling. This could be denoted by grouping all j belonging to one shell in parentheses and adding a common j_{group} in the lower right index:

$$\left[n_i l_i \pm\right]_{j_i}^{x_i},$$

where + denotes spin up $\left(m_s = \frac{1}{2}\right)$ and − denotes spin down $\left(m_s = -\frac{1}{2}\right)$. For example, d+ is an equivalent expression for $j = \frac{5}{2}$. This shell can then couple its whole shell's total angular momentum to the "next" outer shell. An important point about this nomenclature is that there could be parts of the orbital configuration with the same n_i and l_i, because they have a different m_s leading to a different j_i. This leads to the following form of term symbols in JJ-coupling:

$$\left(\ldots\left(\left[n_1 l_1 \pm\right]_{j_1}^{x_1}\left[n_2 l_2 \pm\right]_{j_2}^{x_2}\right)_{j1\&2}\ldots\left[n_k l_k \pm\right]_{j_k}^{x_k}\right)_J^{(o)}$$

Hence the indices in equation $\left(\ldots\left(\left[n_1 l_1 \pm\right]_{j_1}^{x_1}\left[n_2 l_2 \pm\right]_{j_2}^{x_2}\right)_{j1\&2}\ldots\left[n_k l_k \pm\right]_{j_k}^{x_k}\right)_J^{(o)}$ are not the same as in, equation because now they count the portions of equivalent electrons in JJ-coupling. One has to bear in mind that the simple version introduced in equation $\left(j_{1,}j_{2,\ldots,}j_N\right)_J^{(o)}$, now appears distributed over whole term,

$$\left(\ldots\left(\left[n_1 l_1 \pm\right]_{j_1}^{x_1}\left[n_2 l_2 \pm\right]_{j_2}^{x_2}\right)_{j1\&2}\ldots\left[n_k l_k \pm\right]_{j_k}^{x_k}\right)_J^{(o)}$$

In this equation the "last" j_{group} (to be exact $j_{1\&\ldots\&m}$) is the result of successively coupling all subshells and therefore all electrons. This is the same as the whole atom's total angular momentum J and finally just "J" is denoted in the lower right index of the JJ-term symbol.

Atoms in External Fields

There are a number of phenomena associated with the atoms in external electromagnetic fields such as Zeeman Effect, Paschen-Back Effect and Stark Effect. The topics elaborated in this chapter will help in gaining a better perspective about the behavior and phenomena related to the atoms in external fields.

Zeeman Effect

It is well known that an atom can be characterized by a unique set of discrete energy states. When excited through heating or electron bombardment in a discharge tube, the atom makes transitions between these quantized energy states and emits light. The emitted light forms a discrete spectrum, reflecting the quantized nature of the energy states or energy levels. In the presence of a magnetic field, these energy levels can shift. This effect is known as the Zeeman effect. The origin of Zeeman effect is the following. In an atomic energy state, an electron orbits around the nucleus of the atom and has a magnetic dipole moment associated with its angular momentum. In a magnetic field, it acquires an additional energy just as a bar magnet does and consequently the original energy level is shifted. The energy shift may be positive, zero, or even negative, depending on the angle between the electron magnetic dipole moment and the field.

Due to Zeeman effect, some degenerate energy levels will split into several nondegenerate energy levels with different energies. This allows for new transitions which can be observed as new spectral lines in the atomic spectrum. In this experiment we will study Zeeman effect in neon and mercury for which the theory of Zeeman effect is somewhat more tractable.

Non-relativistic quantum theory accounts for only one type of angular momentum called orbital angular momentum. The Hamiltonian for an electron with angular momentum $\vec{\ell}$ has an additional term $\mu_B \vec{\ell} \cdot \vec{H}$ when a weak uniform magnetic field $\vec{H}$ is turned on. μ_B is a constant called the Bohr magneton. First order perturbation theory tells us that energy levels are shifted by,

$$\Delta E = \mu_{\mathrm{B}} M_\ell H$$

where M_ℓ is the quantum number for the component of $\vec{\ell}$ along the field. If an atom had only a single electron and the electron had only "orbital" angular momentum, then equation above would represent the Zeeman shift.

But electrons also have a different type of angular momentum called intrinsic spin angular momentum $\vec{s}$. Spin emerges naturally only in relativistic quantum theories, but it can be shown that inserting a term $g_s \mu_B \vec{s} \cdot \vec{H}$ (where $g_s \approx 2$) into the non-relativistic Hamiltonian gives the correct behavior of spin in a weak field. The first-order perturbation theory gives a corresponding energy shift of,

$$\Delta E = \mu_B g_s M_s H$$

The shift is analogous to that due to orbital angular momentum, except for the constant factor of gs. The energy shift for a particular state depends only on M_s (M_ℓ for the orbital case) for that state. If an atom had only a single electron and the electron had only "intrinsic spin" angular momentum, then equation $\Delta E = \mu_B g_s M_s H$ would represent the Zeeman shift.

Atoms typically have many electrons and are characterized by a total angular momentum $\vec{\ }$ which is the sum of all spin and orbital angular momenta. Though more complicated now, the energy shift is usually expressed in a similar looking form to equations $\Delta E = \mu_B M_\ell H$ and $\Delta E = \mu_B g_s M_s H$.

$$\Delta E = \mu_B g M_J H$$

The important difference is that g depends on the particular state of interest, as does M_J.

By observing the spectra of neon and mercury we will be able to experimentally determine the g-factors for certain states. The values will be compared to the theoretical (Landé) g-factors.

Atomic States in Zero Field

Each of the n electrons in an atom has orbital $\vec{\ell}_i$ and spin $\vec{s}_i$ angular momentum. The sum of all these is the total angular momentum of the atom (ignoring the nucleus).

$$\vec{J} \equiv \vec{L} + \vec{S}$$

$$\vec{L} \equiv \sum_{i=1}^{n} \vec{\ell}_i$$

$$\vec{S} \equiv \sum_{i=1}^{n} \vec{s}_i$$

The convention used here is that angular momentum operators are dimensionless. For example $\vec{\ell} \equiv \vec{r} \times \vec{p} / \hbar$.

- Operators are shown in boldface.
- Gaussian units are used.

The atomic Hamiltonian in no field will be labeled H_0. In the absence of external torques on the atom, the total angular momentum $\vec{J}$ is conserved (i.e. $\vec{J}$ commutes with H_0) and energy eigenstates can be constructed which are also eigenstates of $\vec{J}^2$ and J_z, where the direction chosen for the z-axis is arbitrary. Each eigenstate can be labeled by quantum numbers J and M_J, where:

$$< JLM_J \left| \vec{J}^2 \right| JM_J >= J(J+1) \quad \text{and} < JM_J \left| J_z \right| JM_J > = M_J$$

In principal there are terms in the Hamiltonian that represent interactions between the individual angular momenta of the electrons. As a result, the individual angular momenta, as well as $\vec{L}$ and $\vec{S}$, need not be conserved (i.e. need not commute with H_0). However, these interactions are small effects in many atoms. We will employ the usual approximation, called L-S coupling (or Russel-Saunders coupling), which assumes that the individual orbital angular momenta couple to produce a net orbital angular momentum $\vec{L}$ which has a constant magnitude, but non-constant direction. Similarly the individual spins form a net spin $\vec{S}$ which also has a constant magnitude. (This approximation is found to break down for large Z atoms.) Within this approximation, each eigenstate can be constructed with the form $|JLSM_J>$ with energy $E_0(JLS)$ degenerate in M_J, where,

$$< JLSM_J \left| \vec{L}^2 \right| JLSM_J > = L(L+1) \text{ and } < JM_J \left| \vec{S}^2 \right| JM_J > = S(S+1)$$

These states are not eigenstates of S_z or L_z.

Atomic States in Non-zero Field: Zeeman Effect

Now we will outline how equation $\Delta E = \mu B \text{ g } M_J \text{ H}$ for the Zeeman energy shift can be derived. Within the L-S coupling model, the atomic Hamiltonian in a weak, uniform magnetic field is:

$$H = H_0 + \mu_B \left(\vec{L} + g_s \vec{S}\right) \cdot \vec{H}$$

where μ_B is called the Bohr magneton. It is usually most convenient to choose the z-axis so that $\vec{H} = H\hat{z}$. First order perturbation theory gives energy shifts:

$$\Delta E\left(JLSM_J\right) = \mu_B < JLSM_J \left| L_z + g_s S_z \right| JLSM_J > H$$

If the new term in the Hamiltonian were simply proportional to $J_z = L_z + S_z$ rather than $L_z + g_s S_z$, then the energy shift would be simple and exactly analogous to equation. That is not the case however. The matrix elements in equation $\Delta E\left(JLSM_J\right) = \mu_B < JLSM_J \left| L_z + g_s S_z \right| JLSM_J > H$ cannot be easily evaluated in their present form since the states are not eigenstates of L_z and S_z. However, it can be shown that:

$$< JLSM_J \left| \vec{L} + g_s \vec{S} \right| JLSM_J > = < JLSM_J \left| g\left(JLS\right) \vec{J} \right| JLSM_J >$$

where g(JLS) is the Landé g-factor for a (JLS) state. With this simplification,

$$\Delta E\left(JLSM_J\right) = \mu_B g\left(JLS\right) < JLSM_J \left| J_z \right| JLSM_J > H = \mu_B g\left(JLS\right) M_J H$$

As a result, each state has energy,

$$E\left(JLSM_J\right) = E_0\left(JLS\right) + \mu_B \, g\left(JLS\right) M_J \, H$$

which has the form of equation $\Delta E = \mu B \, g \, M_J \, H$.

The effect of the Zeeman shifts can be seen experimentally. If a particular transition in the absence of an applied field produces radiation at frequency v_o, then the frequency in the presence of a field will be given by,

$$hv = hv_0 + \mu_B \, g(JLS) \, M_J H - \mu_B \, g(J'L'S') \, M_J' H$$

The primed symbols refer to the lower state and the unprimed to the upper state.

Selection Rules

Conservation laws determine which transitions can occur ("allowed") and which can't ("forbidden"). The allowed transitions are specified by a set of conditions called selection rules. The selection rules for these states are given below without derivation.

$$\Delta \ell = \pm 1$$
$$\Delta L = 0, \pm 1$$
$$\Delta S = 0$$
$$\Delta J = 0, \pm 1$$
$$\Delta M_J = 0, \pm 1$$

$\Delta M_J = \pm 1$ transitions are called σ (transitions, while $\Delta M_J = 0$ transitions are called) transitions.

There are additional conditions as well: (1) π transitions between levels both of which have $M_J = 0$ are forbidden if the sum of the J values of upper and lower states is even and (2) J = 0 to J = 0 transitions are forbidden.

It should be noted that the selection rules above only apply to a type of transition called an electric-dipole transition, which is the dominant type. Only "allowed" electric-dipole transitions can occur. However, transitions which are "forbidden" by the electric-dipole selection rules may still take place as other types of transitions.

Landé g-factor

A simple argument can be given for the value of g(JLS). Within the L-S coupling approximation, $\vec{L}$ and $\vec{S}$ are assumed to be individually conserved in magnitude but not direction. Their components parallel to $\vec{J}$ must add to a constant value, but their components perpendicular to $\vec{J}$ are constantly fluctuating. This means that the only part of $\vec{L}$ that contributes to the Zeeman effect is its component along $\vec{J}$, namely:

$$\frac{<\vec{L} \cdot \vec{J}>}{<\vec{J}^2>} \vec{J}$$

$<O>$ stands for $<JLSM_J|O|JLSM_J>$.

Similarly, the only part of $\vec{S}$ that contributes is,

$$\frac{<\vec{s}\cdot\vec{J}>}{<\vec{J}^2>}\vec{J}$$

Then the energy shift, equation $\Delta E(JLSM_J) = \mu_B < JLSM_J|L_z + g_s S_z|JLSM_J > H$ can be replaced by,

$$\Delta E = \mu_B \frac{<\vec{L}\cdot\vec{J}> + g_0 <\vec{S}\cdot\vec{J}>}{<\vec{J}^2>} M_J H$$

Equation $\Delta E = \mu_B \frac{<\vec{L}\cdot\vec{J}> + g_0 <\vec{S}\cdot\vec{J}>}{<\vec{J}^2>} M_J H$ implies that the Landé g-factor is,

$$g(JLS) = \frac{J(J+1)+L(L+1)-S(S+1)}{2J(J+1)} + g_s \frac{J(J+1)+S(S+1)-L(L+1)}{2J(J+1)}$$

In this experiment, certain transitions of neon and mercury atoms will be studied. This section deals with the application of L-S coupling to the particular states involved. All of the spectral lines under consideration in this experiment correspond to allowed transitions, as far as the selection rules are concerned.

States of Neon (Ne) and Mercury (Hg)

First consider neon. The 10 electrons in a neutral neon atom have a ground-state configuration $1S^2$ $2S^2$ $2P^6$. The n = 1 and 2 shells are closed and as such the atom has S = L = J = 0, making the ground state ${}^1S_0$ in spectroscopic notation.

For a given (JLS) state the notation is ${}^{2S+1}X_J$, where X=S means L=0, X=P means L=1, X=D means L=2, etc.

The transitions to be studied in neon are between initial states with one electron excited to a 3P level and final states with one electron excited to a 3S level (not transitions to the ground state). These transitions are simple to study theoretically, because the neon atom can then be treated as a pair of particles — a hole in the n = 2 shell and an electron in the n = 3 shell. In this manner, all 9 unexcited electrons are treated as a single particle, a hole. If we label the excited electron as particle 1 and the hole as particle 2, the upper level for the transitions ($2P^5$ $3P^1$) has $\ell_1 = 1$, $s_1 = 1/2$, $\ell_2 = 1$, and $s_2 = 1/2$. The lower level for the transitions ($2P^5$ $3S^1$) has $\ell_1 = 0$, $s_1 = 1/2$, $\ell_2 = 1$, and $s_2 = 1/2$.

Now consider mercury. The electrons have a ground-state configuration $1S^2$ $2S^2$ $2P^6$ $3S^2$ ... $6S^2$. The transitions to be studied are between states with one excited electron. The initial and final configurations are $6S^1$ $7S^1$ and $6S^1$ $6P^1$, respectively. These mercury transitions are similar to neon in that the angular momentum involves just two particles. If we label the excited electron as particle 1 and the other electron as particle 2, the upper level for the transitions ($6S^1$ $7S^1$) has $\ell_1 = 0$, $s_1 = 1/2$, $\ell_2 = 0$, and $s_2 = 1/2$. The lower level for the transitions ($6S^1$ $6P^1$) has $\ell_1 = 1$, $s_1 = 1/2$, $\ell_2 = 0$, and $s_2 = 1/2$.

For these two-particle states the total orbital angular momentum $\vec{L} \equiv \sum_{i=1}^{n} \vec{\ell}_i$ is simply,

$$\vec{L} = \vec{\ell}_1 + \vec{\ell}_2.$$

This is an operator equation. In terms of the eigenvalues it results in the triangle condition,

$$\left|\vec{\ell}_1 + \vec{\ell}_2\right| \leq L \leq \vec{\ell}_1 + \vec{\ell}_2$$

Thus, for the neon $2P^5\,3P^1$ configuration, the possible values of L are 0, 1, and 2, resulting in S, P, and D states. The $2P^5\,3S^1$ configuration can only lead to a P state. For mercury, the $6S^1\,7S^1$ configuration leads to an S state, while $6S^1 6P^1$ leads to a P state.

In L-S coupling, the total spin $\vec{s} \equiv \sum_{i=1}^{n} \vec{s}_i$ is,

$$\vec{S} = \vec{s}_1 + \vec{s}_2$$

with triangle rule,

$$|s_1 - s_2| \leq S \leq s_1 + s_2$$

Therefore, all states under consideration have S = 0 or 1, giving rise to singlet and triplet states. Finally, L and S are coupled to J with a final triangle rule:

$$|L - S| \leq J \leq L + S.$$

Polarization of the Emitted Light

When an atom undergoes π transition ($\Delta M_J = 0$), its angular momentum about the z-axis does not change. The atom satisfies this requirement by having its optically active electron oscillate along the z-axis, thereby giving rise to an electric field polarized in this direction.

On the other hand, when the atom undergoes a σ transition ($\Delta M_J = \pm 1$), its optically active electron performs a rotary motion in the x-y plane in order that the photon emitted carry angular momentum about the z-axis. The electric field then lies predominately in the x-y plane. Seen edge on, this constitutes a linear polarization perpendicular to the z-axis.

Paschen-Back Effect

In the presence of an external magnetic field, the energy levels of atoms are split. This splitting is described well by the Zeeman effect if the splitting is small compared to the energy difference between the unperturbed levels, i.e., for sufficiently weak magnetic fields. This can be visualized with the help of a vector model of total angular momentum. If the magnetic field is large enough, it disrupts the coupling between the orbital and spin angular momenta, resulting in a different pattern of splitting. This effect is called the Paschen-Back effect.

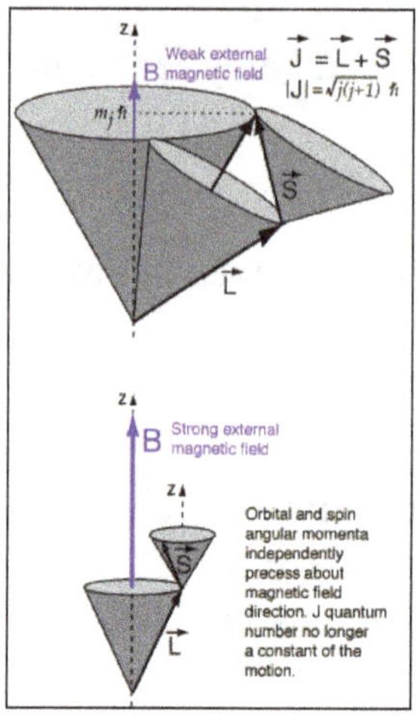

In the weak field case the vector model at left implies that the coupling of the orbital angular momentum L to the spin angular momentum S is stronger than their coupling to the external field. In this case where spin-orbit coupling is dominant, they can be visualized as combining to form a total angular momentum J which then precesses about the magnetic field direction.

In the strong-field case, S and L couple more strongly to the external magnetic field than to each other, and can be visualized as independently precessing about the external field direction.

For reference, the sodium Zeeman effect is reproduced below to show the nature of the magnetic interaction for weak external magnetic fields.

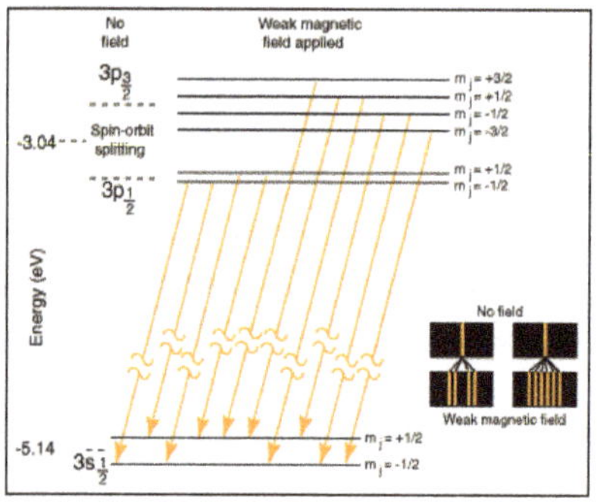

The following is a model of the changes in the pattern if the magnetic field were strong enough to decouple L and S. The resulting spectrum would be a triplet with the center line twice the intensity of the outer lines.

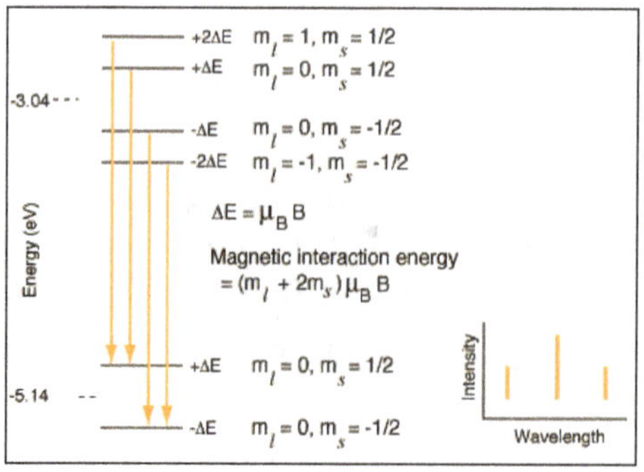

To create this pattern, the projections of L and S in the z-direction have been treated independently and the ms multiplied by the spin g-factor. The energy shift is expressed as a multiple of the Bohr magneton μ_B. The selection rules explain why the transitions shown are allowed and others not.

Sodium was used as the basis of the model for convenience, but the fields required to create Paschen-Back conditions for sodium are unrealistically high. Lithium, on the other hand, has a spin-orbit splitting of only 0.00004 eV compared to 0.0021 eV for sodium. Such small energy values are sometimes expressed in "waven umbers", or $1/\lambda$ in cm^{-1}. In these units the lithium separation is about 0.3 cm^{-1} and the sodium separation is about 17 cm^{-1}. The Paschen-Back conditions are met in some lithium spectra observed on the Sun, so this effect does have astronomical significance.

Stark Effect

The Stark effect is the shift in atomic energy levels caused by an external electric field. There are various regimes to consider. The one treated here is the so-called strong field case, where the shift in energy levels due to the external electric field is large compared to fine structure (although still small compared to the spacings between the unperturbed atomic levels.) In the strong field limit, the Stark effect is independent of electron spin. We start with the ordinary hydrogen Hamiltonian,

$$H_0 = \frac{p^2}{2m} - \frac{e^2}{r}$$

and add a term arising from a uniform electric field along the z- axis.

$$H' = eEz.$$

The + sign on this term. It is easily checked by remembering that the force on the electron due to this term would be obtained by taking $-\sigma_z$, which gives a force along the $-z$ axis, as it should for an electron. To understand the matrix elements that are non-zero, it is useful to temporarily give the external electric field an arbitrary direction,

$$H' = e\vec{\varepsilon} \cdot \vec{x}$$

The selection rules on the matrix elements of $\vec{x}$ are:

$$< n', l', m' \left| \vec{x} \right| n, l, m > \neq 0, \; l' = l \pm 1.$$

These follow from angular momentum conservation ($\vec{x}$ has angular momentum 1), and parity ($\vec{x}$ is odd under parity). Returning to the case of the electric field along the z axis, we have an additional selection rule on m,

$$< n'l', m' \left| z \right| n, l, m > \neq 0, \; l' = l \pm 1, \; m' = m$$

From these selection rules we see that non-zero matrix elements require different values of l. Now for $n=1$ there is only $l=0$,

so, $n=1, l=0 \rightarrow$ no first order Stark Effect.

However, for n = 2, we have two values,

$$n=2, l=0,1, \rightarrow l=0 \leftrightarrow l=1$$

Listing the states for n = 2, we have:

$$\Psi_{200}, \Psi_{211}, \Psi_{210}, \Psi_{21-1}$$

However, l = o has only m = o, so the selection rule on m says we only have non-zero matrix elements,

$$l = 0 \leftrightarrow l = 1 \rightarrow \Psi_{200} \leftrightarrow \Psi_{210}$$

Without the electric field these states have the same energy, so we have a 2 × 2 problem in degenerate perturbation theory. The matrix element we need is proportional to,

$$<210|z|200>=<200|z|210>$$

We label the two degenerate states as follows:

$$200 \rightarrow 1, 210 \rightarrow 2$$

Using this notation, we need to find new linear combinations of these degenerate states, and along the way we find values for the perturbed energy eigenvalues. The equations which accomplish both of these tasks are:

$$\begin{pmatrix} E^{(1)} & 0 \\ 0 & E^{(1)} \end{pmatrix} \begin{pmatrix} c_1 \\ c_2 \end{pmatrix} = \begin{pmatrix} H'_{11} & H'_{12} \\ H'_{21} & H'_{22} \end{pmatrix} \begin{pmatrix} c_1 \\ c_2 \end{pmatrix}.$$

By our selection rules, the diagonal matrix elements vanish,

$$H'_{11} = H'_{22} = 0$$

and

$$H'_{12} = H'_{21} = e\varepsilon <200|z|210>.$$

The wave functions of our states are:

$$\Psi_{200} = N_{20}\left(1 - \frac{r}{2a_0}\right)\exp\left(-\frac{r}{2a}\right)\frac{1}{\sqrt{4\pi}},$$

$$\Psi_{210} = N_{21}\, r \exp\left(-\frac{r}{2a}\right)\sqrt{\frac{3}{4\pi}}\cos\theta,$$

where the N′ s are normalization factors, given below. It is useful at this point to go over to atomic units,

$$z \to a_0 r' \cos\theta,\ \varepsilon \to \frac{e}{a_0^2}\varepsilon',$$

and we subsequently drop the (′) on atomic unit quantities. The matrix element we need is an exercise in elementary integrations. We have,

$$< 200|r\cos\theta|210 >= N_{20}N_{21} \int d\Omega \int r^2\, dr\left(1-\frac{r}{2}\right) \exp(-r/2)r(r\ \exp(-r/2))\ \frac{\sqrt{3}}{4\pi}\ \cos^2\theta = -3,$$

so,

$$H'_{12} = H'_{21} = -3\varepsilon$$

Moving the matrix elements of H′ to the left side of our equation, we have:

$$\begin{pmatrix} E^{(1)} & 3\varepsilon \\ 3\varepsilon & E^{(1)} \end{pmatrix}\begin{pmatrix} c_1 \\ c_2 \end{pmatrix} = 0.$$

This only allows a non zero solution for c_1 and c_2 if the determinant of the coefficients vanishes, or

$$\left(E^{(1)}\right)^2 - (3\varepsilon)^2 = 0.$$

This gives the two eigenvalues,

$$E^{(1)} = \pm 3\varepsilon.$$

These two values determine the amounts by which the n = 2 level is split by the external electric field. To determine the corresponding wave functions, we go back to the equations, with $E^{(1)}$ set equal to one of the eigenvalues. Taking first the lower eigenvalue, we have:

$$E^{(1)} = -3\varepsilon, \begin{pmatrix} -3\varepsilon & 3\varepsilon \\ 3\varepsilon & -3\varepsilon \end{pmatrix}\begin{pmatrix} c_1 \\ c_2 \end{pmatrix} = 0.$$

This gives,

$$c_1 = c_2,\ \Psi_- = \frac{1}{\sqrt{2}}\left[\Psi_{200} + \Psi_{210}\right],$$

where Ψ_- is now the correct unperturbed wave function corresponding to the lower eigenvalue.

Said another way, the perturbing Hamiltonian has matrix element −3E in the state Ψ_-. Doing the same for the upper eigenvalue, we have:

$$E^{(1)} = 3\varepsilon, \begin{pmatrix} 3\varepsilon & 3\varepsilon \\ 3\varepsilon & 3\varepsilon \end{pmatrix} \begin{pmatrix} c_1 \\ c_2 \end{pmatrix} = 0.$$

which gives,

$$c_1 = -c_2, \ \Psi_+ = \frac{1}{\sqrt{2}} \left[\Psi_{200} - \Psi_{210}\right]$$

and Ψ_+ has +3E for the matrix element of the perturbed Hamiltonian. In writing Ψ_+ and Ψ_-, we have supplied the normalization factor $1/\sqrt{2}$. In atomic units, the states Ψ_{200} and Ψ_{210} with normalization factors supplied are,

$$\Psi_{200} = R_{20}Y_0^0 = \frac{1}{\sqrt{2}}\left(1 - \frac{r}{2}\right) \exp\left(-\frac{r}{2}\right) \frac{1}{\sqrt{4\pi}},$$

and

$$\Psi_{210} = R_{21}Y_0^1 = \frac{1}{\sqrt{6}} \, r \, \exp\left(-\frac{r}{2}\right) \sqrt{\frac{3}{4\pi}} \cos\theta$$

with these formulas the wave functions Ψ_+ and Ψ_- are easily constructed and their properties explored. The basic qualitative feature is that Ψ_- favors an increased electron density for $z < 0$ and a decreased electron density for $z > 0$ as expected for the effect of a force directed along the negative z-axis.

Permissions

All chapters in this book are published with permission under the Creative Commons Attribution Share Alike License or equivalent. Every chapter published in this book has been scrutinized by our experts. Their significance has been extensively debated. The topics covered herein carry significant information for a comprehensive understanding. They may even be implemented as practical applications or may be referred to as a beginning point for further studies.

We would like to thank the editorial team for lending their expertise to make the book truly unique. They have played a crucial role in the development of this book. Without their invaluable contributions this book wouldn't have been possible. They have made vital efforts to compile up to date information on the varied aspects of this subject to make this book a valuable addition to the collection of many professionals and students.

This book was conceptualized with the vision of imparting up-to-date and integrated information in this field. To ensure the same, a matchless editorial board was set up. Every individual on the board went through rigorous rounds of assessment to prove their worth. After which they invested a large part of their time researching and compiling the most relevant data for our readers.

The editorial board has been involved in producing this book since its inception. They have spent rigorous hours researching and exploring the diverse topics which have resulted in the successful publishing of this book. They have passed on their knowledge of decades through this book. To expedite this challenging task, the publisher supported the team at every step. A small team of assistant editors was also appointed to further simplify the editing procedure and attain best results for the readers.

Apart from the editorial board, the designing team has also invested a significant amount of their time in understanding the subject and creating the most relevant covers. They scrutinized every image to scout for the most suitable representation of the subject and create an appropriate cover for the book.

The publishing team has been an ardent support to the editorial, designing and production team. Their endless efforts to recruit the best for this project, has resulted in the accomplishment of this book. They are a veteran in the field of academics and their pool of knowledge is as vast as their experience in printing. Their expertise and guidance has proved useful at every step. Their uncompromising quality standards have made this book an exceptional effort. Their encouragement from time to time has been an inspiration for everyone.

The publisher and the editorial board hope that this book will prove to be a valuable piece of knowledge for students, practitioners and scholars across the globe.

Index

A

Alkali Spectra, 100, 124, 132-133, 138

Angular Momentum, 4, 33, 43, 45, 54-56, 62, 65-66, 69, 73-74, 81, 91, 97-98, 132-133, 135, 138, 156, 158, 165-169, 171-176, 178-181

Antideuteron, 57

Atomic Absorption Spectroscopy, 102-103, 105, 111, 113, 115-117

Atomic Bonds, 6

Atomic Mass, 2-3, 10, 17, 50, 59, 117

Atomic Nuclei, 19-21, 23-24, 59

Atomic Orbitals, 36, 38-39, 41, 72-73, 75, 78

Atomic Theory, 13, 25, 27, 62

Aufbau Principle, 39, 41-42, 44, 73

Azimuthal Angle, 33, 89-90

B

Binding Energy, 11, 36, 90, 151

Bohr Atom, 4-5

C

Carbon Atoms, 7-8

Cathode Ray, 28

Chlorine Atom, 7

Chromium, 41, 114, 117

Coulomb Force, 59, 61

Coulomb Potential, 64-65, 84

Covalent Bond, 7-8

D

Deuterium, 2, 48-59, 61, 79

Diatomic Molecule, 13

Doppler Effect, 115

E

Electric Dipole, 56, 154-157

Electromagnetic Radiation, 5, 18, 22-23, 31, 71, 100, 120-121

Electromagnetic Wave, 14

Electron Antineutrino, 57

Electron Configuration, 39, 41-43, 45, 47, 98, 165

Electron Shells, 5, 10, 69, 160-162

Elemental Hydrogen, 48-49

Energy Gap, 41-42, 161

F

Fermions, 4, 9, 19, 54

Fine Structure, 33, 35, 132-133, 135, 137-139, 146-148, 151, 163, 181

G

Gamma Ray, 10, 18, 22-23

H

Heisenberg Uncertainty Principle, 32 36

Helium Atom, 4, 11, 94, 96, 165

Hydrogen Atom, 14, 19, 35, 48, 50, 61, 63-64, 66, 71-72, 90, 94, 122-123, 126-129, 151

Hydrogen Spectrum, 53, 71

Hyperfine Structure, 148-149

I

Ionic Bond, 6-8

Ionization Suppressor, 111

Isotopes, 2, 11, 17, 21, 23, 27, 48-50, 52-53, 149

K

Kinetic Energy, 12, 22, 57, 119, 121

L

Laguerre Polynomials, 70, 76, 78

Law Of Constant Proportions, 26-27

Law Of Reciprocal Proportions, 27

Lithium, 6, 58-59, 61, 114, 117, 124-125, 131, 139, 181

Lorentz Effect, 115

M

Magnetic Field, 4, 9, 15, 18, 23, 32, 38, 133-134, 138, 149-150, 174, 176, 179-180

Magnetic Quantum Number, 39-40, 43, 46, 73-74

Multi Electron Atoms, 80

N

Neon, 6, 17, 39, 117, 174-175, 178-179

Neutron, 2-4, 9-12, 20-23, 50-54, 56, 58, 61

Nuclear Forces, 9, 22

Nuclear Fusion, 12, 48, 50, 53, 57, 59, 61

Nuclear Magneton Units, 56

O

Orbital Quantum Number, 33, 35, 40

P

Pauli Exclusion Principle, 4, 9, 39, 41-43, 85, 171-172

Photons, 5, 10, 20, 24, 93, 120, 122-123, 155, 159-162

Plum Pudding Model, 28, 30

Polar Covalent Bond, 8

Potential Energy, 5, 98-99, 128, 139-141

Probability Density, 37-38, 84-87, 89-90, 126

Protium, 2, 48-51, 53, 57

Q

Quantum Defect, 124-126

Quantum Leap, 5, 11

Quantum Numbers, 33, 39-43, 73-74, 85, 91, 96-97, 133, 167-168, 170-172, 175

R

Radial Probability, 37, 81, 126

Radial Probability Distribution, 81, 126

Radioactive Decay, 11, 57

Rutherford Model, 29-30

S

Sodium Chloride, 7

Sommerfeld Atom Model, 33, 35

Spin Quantum Number, 39-40, 43, 74, 97, 149, 166

Spinning Electron, 4, 138

Stark Effect, 32, 36, 174, 181-182

T

Taylor Expansion, 68, 141

Thomson Atomic Model, 28-29

Tritium, 2, 48-49, 51, 53, 57-61, 79

U

Uranium, 2, 11-12, 17-18, 47, 61, 117

V

Valence Electron, 7, 125-126, 132

W

Wave Equation, 36

Wave Function, 5, 36-37, 54-56, 67-69, 75-76, 84, 86, 89, 135, 165, 183

www.ingramcontent.com/pod-product-compliance
Lightning Source LLC
Chambersburg PA
CBHW082011190326
41458CB00010B/3154
* 9 7 8 1 6 8 2 8 5 9 6 0 5 *